Causes of Climate

Frontispiece The warming trend experienced in the northern hemisphere during the early part of this century has led to glacier retreat in most areas. The illustration shows the Alpine Glacier de la Brenva, in 1767 and 1966. The sketch (above) by Jalabert is remarkably accurate, showing an advanced glacier position in 1767. The photograph (below) from 1966 shows the massive Little Ice Age moraine. *Reproduced by permission of E. Le Roy Ladurie.*

Causes of Climate

John G. Lockwood

A Halsted Press Book

JOHN WILEY & SONS
New York

© John G. Lockwood 1979

First published 1979 by
Edward Arnold (Publishers) Ltd, London WC1B 3DQ

Published in the U.S.A. in 1979
by Halsted Press, A Division of John Wiley & Sons, Inc.,
NEW YORK

Library of Congress Cataloging in Publication Data

Lockwood, John George.
  Causes of climate.

  "A Halsted Press book."
  Bibliography: p.
  1. Climatology. I. Title.
QC981.L62   1978   551.6   78–31907

ISBN 0–470–26657–0
ISBN 0–470–26658–9 pbk.

Printed in Great Britain

Contents

Preface

Causes of Climate has been written for first and second year university, polytechnic and college students in geography and environmental sciences. It is intended to provide arts and social science students with an introduction to modern climatology, and science students with a basis for further advanced study. There is a growing interest in climatic change, and the search for causes of climatic variations on all time-scales now forms a major part of climatology. This book therefore considers the various explanations of climatic change in detail, though it is not meant to be a history of climate. Since explanations of long- and medium-term climatic change depend on modifications or variations in the solar energy reaching the earth's surface, considerable amount of space is given to radiation theory. To many students this may seem an unusual and unnecessary excursion into physics, but it is necessary if the physical bases of modern theories of climatic change are to be understood. As most geography students are required to study elementary mathematics and statistics, simple equations are used in the book where necessary.

I would like to thank the Leverhulme Trust for providing me with a research fellowship for the academic session 1976–7, which enabled me to write this book during my sabbatical year. Dr B. W. Atkinson read the original typescript and made many useful and helpful comments. The book was very carefully typed by Mrs M. Hodgson.

October 1978
Leeds

John G. Lockwood

Acknowledgements

The Publisher's thanks are due to the following for permission to reproduce copyright material:

Royal Meteorological Society for figures 4.16, 5.11 and 6.14; Academic Press Inc. N.Y. and R. Fairbridge for figures 4.24 and 6.12; Dr L. Machta and Plenum Press, N.Y. for figure 6.19; M.I.T. Press, Cambridge, Mass. for figures 4.17, 4.19, 4.23 and 4.26; T. M. L. Wigley, T. C. Atkinson and *Nature* for figures 4.21 and 6.11; Routledge & Kegan Paul Ltd, London, and Elsevier Inc. for figure 2.2; World Meteorological Organization, Geneva for figures 1.1, 2.3, 3.9, 5.3, 5.4, 5.9, 6.18 and 7.1; American Meteorological Society, Boston, Mass. for figures 2.6, 4.4 to 4.7, 4.9, 4.10, 5.16 to 5.18, 6.10 and 6.20; Dr K. G. Bauer for figure 3.13; National Academy of Science for figures 5.1, 5.2 and 5.7; Professor S. Manabe and American Meteorological Society for figure 5.13; *Science* for figures 3.11, 3.12, 5.14, 5.15, 6.17 and 7.2, © 1971, 1975, 1976 The American Association for the Advancement of Science, Washington D.C.; The University of Chicago Press for figure 2.1; Yale University Press, New Haven for figures 5.6 and 5.8; American Geophysical Union, Washington D.C. for figures 4.3 and 5.5; H.U. Roll and Academic Press, N.Y. for figure 3.4; Westdeutscher Verlag, Wiesbaden for figure 6.15; Plenum Publishing Corporation, Boston, Mass. for figure 4.1; International Solar Energy Society, London for figure 2.5; Geological Journal for figure 5.19; HMSO, London for figure 4.15, 4.18 and 5.10; John Wiley & Sons, Inc. N.Y. and Gebrueder Borntraeger, Stuttgart for figure 4.25; The Rockfeller University Press, N.Y. for figure 4.27; Academic Press Inc. (London) Ltd for figure 3.7; Akademie Verlag, Leipzig for figure 4.22; Meteorologische Institut, Bonn and H. Flohn for figures 4.20 and 4.28; and V/0 'Mezhdunarodnaya Kniga', Moscow for figures 4.2 and 4.8.

1
Introduction: The Climatic System

The term climate usually brings to mind an average regime of weather. However, the mechanisms that create the earth's climate and its variations are part of an enormously complex physical system, which includes not only the relatively well known behaviour of the atmosphere, but also the less well known behaviour of the world's ocean and ice masses, together with the variations occurring at the earth's surface. In addition to physical factors there are complex chemical and biological processes affecting the climate, which are also of importance for the climate's impact on living things and thus on man.

The application of both systems theory and mathematics to climatology has changed the subject completely in recent years. The use of systems theory and mathematical models enables the immensely complex real world to be examined and thought about in an orderly, logical manner. Associated with the growth of this systems approach has been a change from the simple description of climatic statistics to a study of the various interactions and exchanges which shape the climatic environment. Classical climatology is concerned with listing averages and extremes of the various climatological elements, but has little to say about the causes of climate. In contrast, modern climatology is concerned with describing the actual causes of climate, and also the causes of any climatic variations. In climatology, the major exchanges consist of flows of energy and water, and much recent research is concerned with tracing the pathways of these particular flows. As mentioned earlier, much use is made of models and the concept of climatological systems, so before discussing basic climatology it is necessary to discuss systems in detail.

1 Systems

A system may be defined as a structural set of objects and/or attributes, where these objects and attributes consist of components or variables that exhibit discernible relationships with one another and operate together as a complex whole, according to some observed pattern. The concept of the system is very useful in providing a means of understanding complex phenomena, provided that it is clearly understood that systems try to describe what happens in nature, and that nature cannot necessarily be forced into the mould of some particular preconceived system. Systems can be classified in terms of their function and also in terms of their internal complexity.

1.1 Isolated, closed and open systems
A common functional division is into isolated, closed and open systems.

(a) Isolated systems have boundaries which are closed to the import and export of both mass and energy. Such systems are rare in the real world, though they may occur in the laboratory, i.e. a mass of gas within a completely sealed and insulated container.

(b) A closed system is one in which there is no exchange of matter between the system and its environment though there is, in general, an exchange of energy. The planet earth together with its atmosphere may, very nearly, be considered a closed system.

(c) An open system is one in which there is an exchange of both matter and energy between the system and its environment. There are numerous examples of open systems in nature, i.e. precipitating clouds, river catchments, plants, etc.

1.1.1 *Isolated systems*
Gas within a completely sealed and insulated container provides a good example of an isolated system. Whatever the original temperature gradients within the gas, temperatures will eventually become uniform, and while the system remains isolated nothing can check or hinder this inevitable levelling down of differences. Stated more generally, in an isolated system there is a tendency for the levelling down of existing differentiation within the system, and towards the progressive destruction of the existing order. In such a system there is always a decrease in the amount of free energy available for causing changes and doing work, and eventually the free energy will become zero.

1.1.2 *Open systems*
Open systems need an energy supply for their maintenance and preservation, and are in effect maintained by the constant supply and removal of material and energy. Closed systems may be considered as a special case of open systems, there being no exchange of matter with the environment. It has already been noted that most of the systems observed within the natural environment belong to the open group. In particular, the open system has one important property which is not found in the isolated system, that is, it may attain a condition known as steady-state equilibrium. This is the condition of an open system wherein its properties are invariant when considered with reference to a given time-scale, but within which its instantaneous condition may oscillate due to the presence of interacting variables. Stated rather more simply, the general features of the system appear to remain constant over a long period of time, though there may be minor changes in details. Meteorological storms, such as hurricanes or thunderstorms are

good examples of open systems in a steady state, in that their general features remain relatively constant over periods of time ranging from several days in the case of the hurricane to several hours for a thunderstorm.

Open systems in the natural environment can be divided into three general categories, which may be termed decaying, cyclic and haphazardly fluctuating. Some systems always belong to one broad category while others change from one to another over relatively short periods of time.

Decaying systems consume their own substance which may be energy or matter, or both. A good example is the decay of river-flow in dry weather, when the flow decreases each day but the rate at which the flow decreases also decreases with time and is proportional to the available water stored in the rocks. The rocks in the river catchment act as a store which supplies water to the river. In this case the river-flow approximates to a negative-exponential decay curve, and the amount of water stored in the rocks decreases to one-half of its original value in a given constant time interval.

The input of short-wave radiation follows diurnal and annual cycles, and these are imposed on many natural systems to form cyclic systems. Heat balances of land surfaces are largely controlled by the input of solar energy, and therefore show both diurnal and annual cycles. Air temperatures reflect the state of the heat balance of the surface and therefore also show marked diurnal and annual cycles. The variations in many cyclic systems when observed over a period of time appear to approximate to a mathematical curve known as a sine curve, which may be obtained by plotting the sine of an angle against the angle itself.

Haphazardly fluctuating systems change in a random and irregular manner, fluctuations occurring at unpredictable times and by unpredictable amounts. Turbulence in fluids or the occurrence of earthquakes are good examples, since neither can be exactly predicted. On small space- and time-scales most systems exhibit some degree of unpredictability.

Systems may also be classified according to their internal complexity, and one frequent type of open system which is found in nature is the cascading system. Cascading systems are composed of a chain of subsystems, having both magnitude and geographical location, which are dynamically linked by a cascade of mass or energy and in this way, the output of mass or energy from one subsystem becomes the input for the next subsystem. Typically, subsystems consist of an input into a store, which may contain a regulator controlling the amount of mass or energy remaining in the store or forming the output. The regulator may be a physical property of the store itself or it may be completely external to the store. More complex subsystems may have several inputs and outputs and even several regulators which decide how the mass or energy is divided between the various outputs. Many of the processes taking place in the atmosphere and the natural environment in general can be interpreted in terms of cascading systems, a good example being provided by the cycle of water. Water may be stored in the oceans, the atmosphere (as water vapour), the soil, the deep rocks, rivers, etc., and the transfer of water from one store to another is controlled by various physical regulators. The output from the atmospheric store in the form of rain constitutes the input into the soil, where in turn one of the outputs forms the input into the deep rock storage, and so on until the water arrives back into the ocean where evaporation forms the input into the atmospheric store.

Interception of rainfall by a forest is a good example of a subsystem. The amount of water that can be carried on a leaf surface is limited, and so there is a definite upper limit to the amount of water than can be stored in a tree canopy and thus to the store of the subsystem. The input into the subsystem is rainfall and the outputs are the evaporation of the intercepted water and the gradual drip of water out of the trees onto the soil surface. At the start of the rainfall the tree canopies will be dry and no water will reach the soil, but after some time the canopies will become completely saturated with water, and when this occurs most of the succeeding rainfall will eventually drip onto the soil surface. So the regulator controlling the amount of water reaching the soil will be the physical geometry of the tree canopies and the percentage saturation of the canopies. There is also a loss of water by evaporation from the intercepted water in the canopies. This loss is controlled by the prevailing meteorological conditions and thus by a regulator which is outside of the physical bounds of the subsystem.

2 The climatic system

From a climatological viewpoint, the atmosphere, oceans and land surfaces may be considered as consisting of a series of cascading systems connected by flows of mass or energy. The hydrological cycle is a good example of such a cascading system, and is discussed in detail in Chapter 4 and illustrated in Figure 4.29. It describes the circulation of water from the oceans, through the atmosphere and back to the oceans, or to the land and thence to the oceans again by overland and subterranean routes. On a global scale it is possible to recognize four major water reservoirs which are the world oceans, polar ice, terrestrial waters, and atmospheric waters. Water in the oceans evaporates under the influence of solar radiation and the resulting clouds of water vapour are transported by the atmospheric circulation to the land areas, where precipitation may occur and the resulting liquid water flow back to the oceans under the influence of gravity. Thus there are two main energy inputs driving the hydrological cycle, and these are gravity and solar radiation. Gravity causes liquid water to run downhill, and under its influence all water would eventually accumulate in the oceans. Solar radiation by causing evaporation, lifts water as water vapour into the atmosphere against the force of gravity. Since this water vapour condenses over the land masses and runs back to the sea, solar radiation allows the hydrological cycle to continue to function. The hydrological cycle is thus driven by a continuous supply of energy from the sun and would soon cease without this particular energy input.

In the same manner that gravity causes water to accumulate in the oceans and leave the lands dry, so friction slows and eventually destroys all motions in the atmosphere and oceans. In the real atmosphere the energy of motion destroyed by frictional forces is replaced by fresh energy derived from solar radiation. So a simple cascade of energy can be recognized in the atmosphere and oceans, since solar radiation is converted into heat which in turn is converted into motion. Motion is destroyed by friction which turns it back into heat, which is lost to space as infrared radiation.

The climatic system consists of those properties and processes that are responsible for the climate and its variations and are illustrated in Figure 1.1. The US National Academy of Sciences (1975) suggests that the properties of the climatic system may be broadly classified as: thermal properties, which include the temperatures of the air, water, ice, and land; kinetic properties, which includes the wind and ocean currents, together with the associated vertical motions, and the motion of ice masses; aqueous properties, which include the air's moisture or humidity, the cloudiness and cloud water content, groundwater, lake levels, and the water content of snow and of land and sea ice; and static properties, which include the pressure and density of the atmosphere and ocean, the composition of the air, the oceanic salinity, and the geometric boundaries and physical constants of the system. These variables are interconnected by the various physical processes occurring within the system, such as precipitation and evaporation, radiation, and the transfer of energy by advection and turbulence.

According to the US National Academy of Sciences the complete climatic system consists of five physical components—the atmosphere, hydrosphere, cryosphere, lithosphere and biosphere. Briefly, they may be described as follows:

Atmosphere This comprises the earth's gaseous envelope, and is the most variable part of the system. The atmosphere, by transferring heat vertically and horizontally, adjusts itself to an imposed temperature change in about a month's time. This is also approximately the time it will take for the atmosphere's kinetic energy to be dissipated by friction, if there were no processes acting to replenish this energy.

Hydrosphere This comprises the liquid water distributed over the surface of the earth, including the oceans, lakes, rivers and groundwater. Of these, the world's oceans are the most important for climatic variations. The ocean absorbs most of the solar radiation that reaches the earth's surface, and also represents an enormous reservoir of heat due to the relatively large mass and specific heat of the ocean's water.

Cryosphere This comprises the world's ice masses and snow deposits, which includes the continental ice sheets, mountain glaciers, sea ice, surface snow cover, and lake and river ice. Changes in snow cover are mainly seasonal in character, while glaciers and ice sheets respond much more slowly. Glaciers and ice sheets only show significant changes in volume and extent over periods ranging from hundreds to millions of years.

Lithosphere This consists of the land masses over the surface of the earth, and includes the mountain and ocean basins, together with the surface rock, sediments, and soil. These features change over the longest time scales of all, the components of the climatic system ranging up to the age of the earth itself.

Biosphere This includes the plant cover on land and in the ocean and the animals of the air, sea and land. Biological elements are sensitive to climate and, in turn, may influence climatic changes.

Primary among the processes responsible for climate is the rate at which heat is added to the climatic system, the ultimate source of which is of course the sun's radiation. The atmosphere and the oceans respond to this heating by developing winds and currents, which in turn serve to transport heat from regions where it is received in abundance to regions where there is a thermal deficit. A great deal of this heat is transported by the large-scale transient disturbances in the atmosphere and ocean. Because of the large thermal capacity of the system compared to the rate of heating, individual large-scale weather systems are, within their lifetimes of about one week, relatively little affected by heating. Beyond a few weeks, on the other hand, the heating becomes essential for re-supplying the energy of the atmospheric system. This is why so much attention is given to radiation and energy supply in this book which is basically about climate.

2.1 **The nature of energy**

Energy may formally be defined as the capacity for doing work, and it may exist in a variety of forms including heat, radiation, potential energy, kinetic energy, chemical energy, and electric and magnetic energies. It is a property of matter capable of being transferred from one place to another, of changing the environment and is itself susceptible to change from one form to another. An example of energy changing the environment is provided by solar radiation falling on a field during the early morning and increasing both the temperature of the air and of the plants. Another example is the energy of high winds or floods which may change the natural environment in a far more spectacular and destructive manner. If nuclear reactions are excluded, it can be stated that energy is neither created nor destroyed, and from this it follows that all forms of energy are exactly convertible to all other forms of energy, though not all transformations are equally likely. It is therefore possible for any particular system to produce an exact energy account, in which the energy gained exactly equals the energy loss plus any change in storage of energy in the system. Since a continuous transformation of energy from one form into another takes place in the atmosphere and on the earth's surface, it is necessary to consider in some detail the various forms which energy can assume.

2.1.1 *Heat*

Heat is a form of energy and it defines in a general way the aggregate internal energy of motion of the atoms and molecules of a body. It may be taken as being equivalent to the specific heat of a body multiplied by its absolute temperature in degrees Kelvin and by its mass, where the specific heat of a substance is the heat required to raise the temperature of a unit mass by one degree. It is important to distinguish between temperature and heat, for temperature is a measure of the mean kinetic energy (speed) per molecule of the molecules in an object, while heat is a measure of the total kinetic energy of all the molecules of that object. As the temperature increases so does the mean kinetic energy of the molecules, and conversely it is possible to imagine a state when the molecules are at complete rest, a point on the temperature scale known as absolute zero. This has been found to be at 273·15 Celsius degrees below the melting point of ice (0 °C), and the Kelvin temperature scale is measured from absolute zero in Celsius units making 0 °C equivalent to 273·15 K. The Kelvin temperature scale is used in basic physical equations which involve temperature and it has the practical advantage of avoiding negative values.

Temperature is the condition which determines the flow of heat from one substance to another, the direction being from high to low temperatures. So long as only one object is considered, its temperature changes represent proportional changes in heat content. The definition of heat content suggests that when a variety of masses and types of material are considered, the equivalence of heat and temperature disappears. Often a small hot object will contain considerably less heat than a large cool one, and even if both have the same mass and temperature their heat contents can differ because of differing specific heats.

The transfer of heat to or from a substance is effected by one or more of the processes of conduction, convection or radiation. The common effect of such a transfer is to alter either the temperature or the state of the substance or both. Thus, a heated body may acquire a higher temperature (sensible heat) or change to a higher state (liquid to gas, or solid to liquid) and therefore acquire latent or hidden heat. Conduction is the process of heat transfer through matter by molecular impact from regions of high temperature to regions of low temperature without the transfer of the matter itself. It is the process by which heat passes through solids but its effects in fluids (liquids and gases) are usually negligible in comparison with those of convection. In contrast, convection is a mode of heat transfer in a fluid, involving the movement of substantial volumes of the substance concerned. Conduction is the main method of heat transfer in the solid rocks and the soil, while the convection process frequently operates in the atmosphere and oceans.

2.1.2 Radiation

This is the transmission of energy by electromagnetic waves, which may be propagated through a substance or through a vacuum at the speed of light. Electromagnetic radiation is divided into various classes which differ only in wavelength; these are, in order of increasing wavelength: gamma radiation, X-rays, ultraviolet radiation, visible radiation, infrared radiation and radio waves. All objects which are not at the absolute zero of temperature give off radiant energy to the surrounding space, so the environment is full of radiation of various wavelengths, the most important of which are in the visible and infrared sections. Furthermore, since nearly all the available energy in the natural environment was originally gained as visible radiation from the sun, the study of radiation is obviously of great importance and is considered separately in greater detail in the next chapter.

For the sun, the wavelength of maximum emission is near 0·5 μm (10^{-6} m), which is in the visible portion of the electromagnetic spectrum, and almost 99 per cent of the sun's radiation is contained in the so called short wavelengths from 0·15 to 4·0 μm. Observations show that 9 per cent of this short-wave radiation is in the ultraviolet (less than 0·4 μm), 45 per cent in the visible (0·4 to 0·7 μm) and 46 per cent in the infrared (greater than 0·74 μm). The surface of the earth, when heated by the absorption of solar radiation, becomes a source of long-wave radiation. The average temperature of the earth's surface is about 285 K, and therefore most of the radiation is emitted in the infrared spectral range from 4 to 50 μm, with a peak near 10 μm. This radiation may be referred to as long-wave, infrared, terrestrial or thermal radiation. Net radiation is the difference between the total incoming and total outgoing radiation, and clearly it shows whether net heating or cooling is taking place.

2.1.3 *Potential energy*
This is the energy possessed by a body by virtue of its position. It is measured by the amount of work required to bring the body from a standard position, where its potential energy is zero, to its present position. Thus a body at some distance above the ground has more gravitational potential energy than a body at ground level, and if released the potential energy will be converted into kinetic energy as the object accelerates towards the earth. Rivers are good examples of the conversion of potential energy into kinetic energy. Water vapour in the atmosphere possesses some gravitational potential energy in respect of its altitude above sea-level, and this potential energy is converted into kinetic energy when it condenses into rain which then falls towards the ground. If the rain reaches sea-level, then all the gravitational potential energy of the rainwater will have been converted into kinetic energy, but in contrast, if it falls on an upland surface, some potential energy will still be available and this will appear as the energy of river-flow as the water moves towards the sea along stream channels.

2.1.4 *Kinetic energy*
This is the energy possessed by a body by virtue of its motion. It is a quantity of magnitude $\frac{1}{2}MV^2$, where M is the mass and V the velocity of the particle. Kinetic energy is continuously dissipated by the various resistances to motion, and is often converted into heat. The kinetic energy of rivers is dissipated by the resistance to water movement created by the uneven stream floor. Since this resistance is very large, rivers normally flow only very slowly, suggesting that kinetic energy is being destroyed almost as fast as it is created from the gravitational potential energy of the water. The atmosphere contains kinetic energy because of the winds, and this is dissipated mainly by friction at the ground surface. It is estimated that, in the absence of solar radiation which creates kinetic energy, dissipation of the atmosphere's kinetic energy by friction would be almost complete after six days. Kinetic energy is therefore, in the natural environment, one of the less

stable and more short-lived forms of energy, and will soon be converted into other forms unless it is continually renewed.

2.1.5 *Chemical energy*
This is the energy used or released in chemical reactions. Some chemical reactions, such as the process of burning or exploding, release large amounts of energy in the form of heat and light, whereas others absorb or release energy only very slowly. In the natural environment there are various chemical reactions, one of the more important being the process of photosynthesis in plant leaves. In photosynthesis the elements of two atmospheric gases (carbon dioxide and water vapour) are combined with light energy captured by the chloroplasts of plant leaves to form plant materials and oxygen:

$$CO_2 + 2H_2O \xrightarrow[\text{green plant}]{\text{light}} (CH_2O) + O_2 + H_2O$$

Therefore plants normally require the energy contained in sunlight to grow. If plant materials are burnt, the energy originally absorbed over long periods of time is released in a short intense burst of heat and light. The amount of photosynthesis which takes place on earth may be estimated from the amount of carbon fixed from carbon dioxide each year. Estimates of this type show that 90 per cent of carbon is fixed by aquatic plants and the remainder by land plants, of which forests fix 7 per cent leaving only 3 per cent for all the managed and unmanaged fields on earth.

2.1.6 *Electric and magnetic energies*
These are of little importance in the natural environment at the earth's surface, but they are of great interest in the very high atmosphere on the edge of space. The most obvious manifestation of electric energy near the surface is provided by lightning in thunderstorms.

2.2 Energy transformations
Continual transformations of energy from one form to another take place in the atmosphere and on the earth's surface. Most of these transformations start with solar radiation and end with the loss of infrared radiation to space. Some examples of simple transformations have been discussed by Lockwood (1976) and are listed below.

A very simple series of transformations may be observed when solar radiation falls on a desert surface:

$$\text{solar radiation} \longrightarrow \text{heat} \longrightarrow \text{infrared radiation}$$

The solar radiation causes an increase in the heat content of the soil and the air, and this is observed as a rise in temperature. Part of the additional heat is eventually lost as infrared radiation to space.

A variation of the first case is:

$$\text{radiation} \longrightarrow \text{(sensible heat + latent heat)}$$

If the radiation falls on a water or moist surface (plants etc.), some of the radiant will be used in warming the surface, forming sensible heat, but some will be used in evaporating water vapour and this will form latent heat. Sensible heat is heat that can be felt and detected by a change in temperature, whereas in contrast, latent heat is completely hidden and is only detected when a change of state of the substance concerned takes place. Thus when water vapour condenses into water droplets the following transformations take place:

$$\text{latent heat} \longrightarrow \text{sensible heat} \longrightarrow \text{infrared radiation}$$

The latent heat taken up at evaporation is released and appears as sensible heat which will eventually form radiation. This process is observed in clouds where the sensible heat released warms the surrounding atmosphere. The reverse process takes place during evaporation:

$$\text{potential energy} \longrightarrow \text{kinetic energy} \longrightarrow \text{heat} \longrightarrow \text{radiation}$$

This sequence is a very common one, since kinetic energy is usually dissipated by friction and is converted into heat, which is lost by infrared radiation to space.

Now all the above transformations of energy can proceed in the reverse as well as the forward direction. Thus kinetic energy can become potential energy, sensible heat can become latent heat and so on, and indeed the only transformation which is not usually observed is the direct conversion of radiant energy into kinetic energy.

2.2.1 *Energy in the atmosphere*
Energy transformations which occur in the natural environment are well illustrated (Lockwood 1976) by a study of the atmosphere where the various components of energy may be found. The energy content of one gram of moist air may be written as follows:

total energy content (Q) = latent heat content (Lq) + sensible heat content
(C_pT) + potential energy content (gz)
+ kinetic energy content $(V^2/2)$

In the following discussion it is assumed that one gram of moist air is under discussion, so mass will often be omitted. The first component of the total energy content is the latent heat term, which consists of the latent of vaporization (L) multiplied by the mass of water vapour (q). Latent heat is formally defined as the quantity of heat absorbed or emitted, without change of temperature, during a change of state of unit mass of a material and for vaporization of water is equivalent to 575 cal g^{-1}. Since one calorie is the amount of heat required to raise the temperature of one gram of water by one degree Celsius, it is seen that a large amount of energy is locked as latent heat in water vapour. Indeed, in moist tropical air the latent heat component can amount to about 15 per cent of the total energy content.

Sensible heat is the product of the specific heat (C_p) and the temperature

(T) in degree K, and is contained in the second term. The atmosphere is continually losing sensible heat, in the form of radiation to space, at a rate of about $\frac{1}{4}$ per cent per day of the total energy content; this represents a cooling of about 1 to 2 degree C per day. Sensible heat is gained from the surface and by the release of latent heat from condensing water vapour.

Potential energy exists in the unit parcel by virtue of its position above the earth's surface. It is the product of the height z above sea-level and the force of gravity g. If the parcel sinks slowly, the potential energy must decrease and re-appear as another form of energy, which normally takes the form of sensible heat. In the atmosphere there is a very close relationship between potential energy and sensible heat, since as air parcels sink, their potential energy is converted into sensible heat and as they rise, their sensible heat is converted back into potential energy.

An adiabatic process is one in which heat does not enter nor leave the system, and so there are no radiative gains or losses, no heat flows, no mixing with the surrounding environment, and no changes in water content. Thus the total energy content of a unit air parcel will remain constant during an adiabatic change. Furthermore, if there is no condensation nor evaporation within the parcel, that is to say, if the latent heat component remains constant, then under adiabatic conditions the sum of the sensible heat and the potential energy will also remain constant, and it is convenient to consider these two quantities together under the heading of total potential energy. Since the total potential energy is constant in an adiabatic process, the temperature of a rising air parcel must decrease since this is the only way in which the sensible heat content can decrease, the converse being true for sinking parcels. This is observed in the atmosphere, and it is found that the temperature change is about 0·98 degree C per 100 m, which is known as the dry adiabatic lapse rate.

Pressure decreases with height in the atmosphere and this can also be used to explain the dry adiabatic lapse rate. If a parcel of air rises, it expands because of the lower environmental pressure, and since the work done by the parcel in so expanding must be at the expense of its internal energy, its temperature falls despite the fact that no heat leaves the parcel. Conversely, the internal energy of a falling parcel is increased and its temperature raised, as a result of the work done on the air in compressing it.

If water vapour is present in the rising air parcel, a stage will be reached when the air becomes saturated and condensation occurs, thus releasing latent heat. The released latent heat becomes sensible heat and therefore decreases the net lowering of the sensible heat content of the rising parcel. The net result is that the temperature change with increasing altitude is less than the dry adiabatic lapse rate, and is at a rate known as the saturated adiabatic lapse rate. Since the rate of condensation of water vapour varies with temperature, the saturated adiabatic lapse rate does not have a fixed value, but varies from near that of the dry adiabatic lapse rate for cold air to considerably less for very warm air.

Observation shows that it is justifiable to treat vertically moving individual masses of air as isolated systems which move through the atmosphere without unduly disturbing it or exchanging heat with it. As widespread vertical motion occurs in the lower atmosphere, the average lapse rate of temperature with height is between the dry and saturated adiabatic lapse rates and averages 0·6 degree C per 100 m. Naturally various non-adiabatic processes such as condensation, evaporation, radiation and turbulent mixing also operate to produce temperature changes in the lower atmosphere, but their effects are generally negligible in comparison with those caused by appreciable vertical motion.

Kinetic energy ($V^2/2$) forms a very small part of the total energy content of the atmosphere, amounting to only about 0·5 per cent of the total in the regions of strongest wind speed and considerably less elsewhere. The bulk of the atmospheric energy content is contained in the form of sensible heat plus potential energy, and in normal calculations the energy of the winds is so small that it can be neglected.

The equation for the total energy content of a unit mass of air can be used to study the energy balance of the tropical atmosphere. Outside of southern Asia, the mean north–south circulation of the tropical atmosphere can be considered as taking the form of two simple cells, with rising air near the equator and sinking air over the subtropical deserts. The low-level circulations of these cells form the north-east and south-east trade winds. Sinking air in the subtropics increases its temperature according to the dry adiabatic lapse rate, thus resulting in clear skies and low relative humidities. Subtropical deserts are largely a result of atmospheric subsidence leading to cloudless and rainless conditions.

Large areas of the subtropics consist of ocean, and the clear skies result in a plentiful supply of solar radiation reaching the surface where it is mostly used to evaporate sea water. Therefore, over the subtropical oceans radiant energy from the sun is turned partly into sensible heat which warms both the atmosphere and ocean, but mostly it is converted into latent heat. There is no evaporation in the dry subtropical deserts, so all the incoming radiation is turned into sensible heat resulting in very high day-time temperatures. Radiation losses are also high in deserts, and at night temperatures fall to low values, so the net gain in the sensible heat content of the atmosphere is often small. Water vapour evaporated over the subtropical oceans is mixed through the lower layers of the atmosphere by turbulence and convection and carried towards the equator by the trade winds. Near the equator, the trade winds enter the equatorial trough and here ascent takes place in localized weather systems and in particular in thunderstorms. In the thunderstorms, the latent heat released by the condensing water vapour is converted into sensible heat which in turn is transformed into potential energy by the rising air mass. In this way, the total potential energy (sensible heat + potential energy) of the rising air in the thundercloud is increased by the release of latent heat, and is then exported at high levels in the atmosphere into the subtropics and also into middle latitudes.

The atmosphere is not therefore warmed directly by solar radiation, for indeed in the subtropics air is actually sinking over the regions of highest surface temperatures, which often decrease towards the equator. There exists instead, a complex mechanism whereby solar energy is turned into latent heat which is transported into the equatorial trough and there converted into sensible heat, so that although the main input of solar energy is in the sub-tropics, the main heating of the atmosphere occurs near the equator. Regions of intense convection with associated clouds in the equatorial atmosphere can be regarded as sites where large amounts of sensible heat are being released and the atmosphere is actually being warmed. It has been calculated that only about 1,500–5,000 active thunderstorms are required to maintain the heat budget of the equatorial trough and thus provide for most of its poleward energy export.

2.3 The general circulation of the atmosphere

The term 'general circulation' in its widest sense is used to imply all aspects of the three-dimensional global flow and energetics of the whole atmosphere. In this sense the general circulation is exceedingly complex, involving fluctuations on all time- and space-scales, and forming part of the climatic system described earlier.

The planet earth receives heat from the sun in the form of short-wave radiation, but it also radiates an equal amount of heat to space in the form of long-wave radiation. This balance of heat gained equalling heat lost only applies to the planet as a whole over several annual periods; it does not apply to any specific area for a short period of time. The equatorial region absorbs more heat than it loses, while the polar regions radiate more heat than they receive. Nevertheless, the equatorial belt does not become warmer during the year, nor do the poles become colder, because heat flows from the warm to the cold regions, thus maintaining the observed temperatures. An exchange of heat is brought about by the motion of the atmosphere and the upper layers of the oceans, thus forming the general circulation of the atmosphere and oceans.

Interest in the general circulation of the atmosphere arises because it is one of the major factors controlling the distribution of world climatic zones. In the tropics vast areas have easterly winds with a large equatorial component, which are the so called trade winds noted for their generally fine, sunny weather. Poleward of the trades there are regions of subsiding air with generally anticyclonic conditions, which form the great deserts of the planet. In middle latitudes the average atmospheric motion is towards the east, but the flow is highly disturbed. Lastly, the polar regions with their high radiation loss have generally subsiding air masses.

Inequalities in heating start the basic north–south motion from warm to cold regions, but the rotation of the earth also generates east–west motions in the atmosphere. Chapter 4 describes how air moving toward the pole is deflected to the right in the northern hemisphere and to the left in the southern. The rotation of the earth also helps to generate large rotating storms which carry out much of the middle-latitude heat exchange. These middle-latitude storms travel mainly from west to east, and form a never-ending succession of low-pressure centres, or depressions, with winds revolving anti-clockwise in the northern hemisphere, followed by high-pressure centres, or anticyclones, with clockwise circulations. The number of travelling storms is highest near latitude 45°. On a planet with a uniform surface, low- and high-pressure centres would march with equal frequency across all longitudes at one latitude. However, because of the distribution of oceans, continents and mountain ranges, high- and low-pressure centres are consistently found over some regions and, just as consistently, not over others. Thus the average flow during any one season contains distinct cellular patterns.

An air mass is a body of air in which the horizontal gradients of temperature and humidity are relatively slight and which is separated from an adjacent body of air by a sharply defined transition zone, known as a front, in which these gradients are relatively large. Horizontally homogeneous bodies of air are produced by prolonged contact with an underlying surface of uniform temperature, known as a source region. Source regions must have light winds and are therefore usually found in the permanent or semi-permanent high-pressure systems—the subtropical, polar and winter continental anticyclones. This leads to a general classification of air masses as 'polar' or 'tropical', maritime or continental, défining their basic temperature and humidity characteristics.

Tropical air originating over the oceans in the subtropical highs around 30°–35° N and S is known as maritime tropical air. It is quite warm and, near the surface, moist; it forms the trade winds and also flows into the westerlies of the temperate latitudes. Continental tropical air forms over the large deserts of the tropics, and it is characteristically extremely hot and dry at the surface, but when it flows over the sea it is gradually modified to form maritime tropical air.

Polar air originates in high latitudes and may be subdivided into maritime polar and continental polar according to the nature of the surface over which it formed. The former is only relatively cool and usually very moist and it feeds into the poleward edge of temperate latitude westerlies. Extensive anticyclones form in winter over the continental interiors of Asia and North America and these are the sources of extremely cold, dry winds which blow out of the continental interiors in winter.

Polar and tropical air masses meet in middle latitudes along a front which is given the general name 'polar front'. The warm tropical air rises over the cold polar air, so the front rises towards the pole with a slope of about 1 in 200. The polar front is usually undisturbed, being marked by a line of cloud and a little light rain, but occasionally a small wave forms on the front, the warm air penetrating horizontally into the cold air. Some of these waves grow rapidly in a period of one or two days, and in particular the penetration of the warm air in the form of the warm sector increases, resulting in a very pronounced wave structure. Much of the ascent and weather in a frontal depression takes place along the warm and cold fronts. At the warm front, warm air displaces cold along a gently sloping surface of about 1 in 200, spreading rain and layer cloud over a wide area ahead of the depression, while at the cold front cold air displaces warm air along a steeply sloping surface of about 1 in 50, generating a narrow belt of cloud and rain.

The initial wave disturbance grows in a period of about a day into the frontal depression shown in Plate 1. As the warm air ascends and escapes to higher levels over the warm-front surface and as the cold front undercuts the warm air, the warm sector narrows, with the result that the cold front tends to overtake the warm front, and the cyclone is said to be occluded. The front resulting from the combination of the warm and cold fronts is called an occlusion. In the early stages of depression development, the lowest pressure is found at the tip of the wave formed in the polar front; but as the

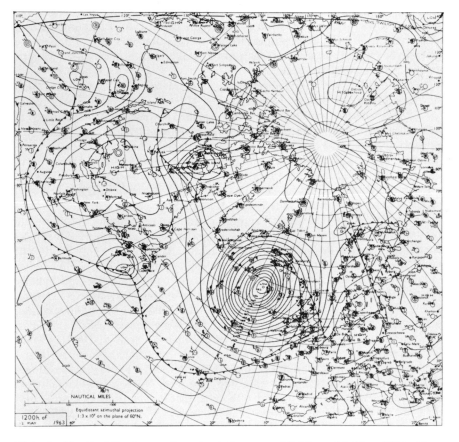

Plate 1 North Atlantic weather chart for noon GMT on 12 May 1963. Deep depression with associated occlusion to the south of Iceland. Frontal depression with a wide warm sector to the south of Newfoundland. *Reproduced by permission of the Controller of Her Majesty's Stationery Office, Crown Copyright reserved.*

occlusion process continues, the fronts gradually become separated from the region of lowest pressure. In the later stages of the life cycle, the occlusion process becomes almost complete, and the warm air is lifted completely off the surface. A frequently occurring feature of the partly-occluded and occluded stages is the development in the rear of the depression of a trough, which is often accompanied by bad weather due to general convergence in the lower troposphere.

In the northern hemisphere, depressions frequently form in winter off the eastern coasts of Asia and North America. During the early stages of development, the cyclones move rapidly north-east, but after 24–30 hours they start to occlude and slow down. They frequently become stationary over the eastern parts of the oceans, and slowly decay over several days. Indeed, it is the frequency with which depressions become stationary near Iceland or the Aleutian Islands which gives rise to the Icelandic and Aleutian lows on mean pressure charts (see Figure 4.18). Furthermore, it is observed that frontal depressions rarely appear alone, for normally there are two, three or more in a series, each in the wake of the other with a general tendency to move north-east.

Frontal depressions are primarily maritime storms which are best developed over the oceans. In the interior of continents, particularly to the east of the Rocky Mountains and in central and eastern Eurasia, the areas of continuous cloud cover and precipitation may be small, and in many cases the precipitation area may be broken or absent. Indeed, the weather distribution within a cyclone depends very much on the location and the time of year, and it is not possible to generalize with safety. For example, cold fronts are often very active with severe thunderstorms over the eastern USA, but they rarely give any significant weather in western Europe.

Anticyclones are the other major weather system found in temperate latitudes. They often appear as sluggish and passive systems which fill the spaces between the far more active depressions. Anticyclones may be divided very broadly into cold or polar continental highs and warm or dynamic highs.

Polar continental highs develop in winter over the northern land masses, where they are caused by the intense cooling of the snow-covered surface, giving rise to a shallow and dense layer of very cold air. The relatively high density of the cold air increases the surface pressure above normal and so an anticyclone appears on the surface pressure chart. These cold anticyclones are therefore only shallow formations and it is rare that they can be traced more than 2 km above sea-level. The most pronounced and persistent example is found over Siberia, the corresponding high over North America being far less regular. Both these highs are, of course, the main sources of continental polar air in the northern winter.

The warm or dynamic anticyclone is caused by large-scale subsidence throughout the depth of the lower atmosphere, and good examples are the highs associated with the descending limbs of the tropical circulation cells. An essential feature of all dynamic anticyclones is a low-level temperature inversion. The general subsidence usually ceases at the top of this inversion and the weather associated with the high often depends on the nature of the comparatively cool air near the surface. If the surface air is moist, cloud may form below the inversion and light rain may even fall, and this is particularly likely to occur where the flow is from a warm sea across a cold land or sea surface. In contrast, when the surface air is relatively dry, the skies may be almost cloudless, leading to great heat in summer and intense cold in winter. As with depressions, it is not possible to generalize about the weather distribution within these systems. Semi-permanent dynamic anticyclones are common in the subtropics, but they can also occur anywhere in the middle latitudes at any season of the year.

In winter, the belt of middle-latitude westerlies broadens, and the high-pressure areas over the subtropics move toward the equator by 5° to 10° latitude. Both the frequency and the intensity of cyclones are greater in winter than in summer. The effect of the continents on seasonal variations is most marked over Asia. In winter, air flows outward in a wide arc from the Siberian anticyclone toward the tropics, but the flow is reversed during the summer. This seasonally reversing circulation is called a monsoon, from an Arabic word denoting a seasonal wind.

The influence of the earth's surface on air flow is often detectable up to an altitude of 2 km. In a deep layer of the atmosphere above the influence of surface friction, westerly winds encircle the globe from latitudes 20° to 25° to the poles. Daily weather charts for any large region in middle and high latitudes reveal well defined weather systems that normally move from west to east with a speed that is considerably smaller than the speed of the upper winds. The structure of these systems varies considerably with altitude, for near the earth's surface they are of relatively small dimensions (1,000 to 3,000 km) and show a maximum of complexity, while in the middle and upper atmosphere the systems are relatively large and have a great simplicity. At the surface the predominant features are closed cyclonic and anticyclonic systems of irregular shape, while higher up smooth wave-shaped patterns are the general rule. The dimensions of these upper waves are much larger than those of the surface cyclones and anticyclones, and only rarely is there a one-to-one correspondence. In typical cases there are four or five major waves around the hemisphere, and superimposed upon these are smaller waves which travel through the slowly moving train of larger waves. The major waves are called long waves or Rossby waves. Troughs are found in winter over eastern North America, over eastern Europe, and near the Asiatic east coast. Ridges lie above the eastern Atlantic and Pacific Oceans, and over central Siberia.

Despite the tendency of long waves to remain in preferred positions, their location and intensity at any time undergoes frequent variations. A narrow, high-velocity core within the upper westerlies, the so-called jet stream, is one of the atmosphere's main links in communicating such changes. Winds frequently reach speeds of 75 m s^{-1} in the jet stream, carrying air around one-fifth of the world in one day at latitudes 40° to 50°.

The general circulation of the atmosphere produces a broad latitudinal zonation of the world climatic zones. Thus a warm zone is found near the equator corresponding to the ascending limb of the tropical circulation cells, but in contrast the descending limbs in the subtropics cause widespread aridity. The travelling disturbances of the middle latitudes create cool, humid zones poleward of the subtropical deserts, while precipitation is low in the polar regions because the air is too cold to hold much water vapour. This very simple climatic zonation is distorted by the continents. The precipitation-forming disturbances of the temperate latitudes rarely penetrate into the continental interiors, which therefore remain dry and tend to be poleward extensions of the subtropical deserts. The largest modification is found over southern Asia where the normal wind circulation is reversed during the northern summer, bringing rain to areas which would normally be subtropical desert.

2.4 Climatic change and its causes

It is universally accepted that global climate has undergone significant variations on a wide variety of time-scales, and there is every reason to expect that such variations will continue in the future. Thus the major climatic events during the past 150,000 years were the occurrence of two glacial maxima of roughly equal intensity, one about 135,000 years BP, and the other between 14,000 and 22,000 years BP. Both were characterized by widespread glaciation and generally colder climates and were abruptly terminated by warm interglacial intervals that lasted about 10,000 years. The penultimate interglacial reached its peak about 124,000 years ago, while the present interglacial had its thermal maximum about 6,000 years ago. In general, the period about 7,000 to 5,000 years BP, was warmer than today, and the past 7,000 years has been punctuated in many parts of the world by colder intervals about every 2,500 years, with the most recent occurring about 300 years ago. The average surface air temperature in the northern hemisphere increased from the 1880's until about 1940 and has been decreasing thereafter. From 1950 to 1975 the rate of cooling of most of the climatic indices in the northern hemisphere was between 0·1 and 0·2 degree C per decade. In 1971–5 the sea surface in the north central Pacific and North Atlantic was significantly colder, snow cover area significantly larger and average temperatures of the atmosphere in the low and middle latitudes significantly lower than in the previous five-year period. Data from the southern hemisphere show cooling until the mid 1960's, and then a slight warming.

The view that climatic variation is a strictly random process in time can no longer be supported. The very concept of a climatic system suggests that many atmospheric variables are correlated on time-scales of weeks, months, and even years. Thus variations of the global ice distribution, for example, have a significant effect on the net heating of the atmosphere, by virtue of the ice's effective control of the surface albedo. The inputs to the climatic system shown in Figure 1.1 may be considered to be the configuration of the earth's crust and the incoming solar radiation. Each of these may be used to develop a climatic theory to explain certain features of the observed climatic changes. For example, changes of the distribution of solar radiation known as the Milankovitch mechanism have been used to explain the major glacial–interglacial cycles of the order of 10,000 to 100,000 years. The Milankovitch mechanism consists of variations of the earth's orbital parameters which produce changes in the intensity and geographical pattern of the seasonal radiation received at the top of the atmosphere and in the length of the radiational seasons in each hemisphere. The orbital parameters (eccentricity, obliquity, and precession) vary with periods averaging about 96,000 years, 41,000 years and 21,000 years, respectively. There is also the separate ques-

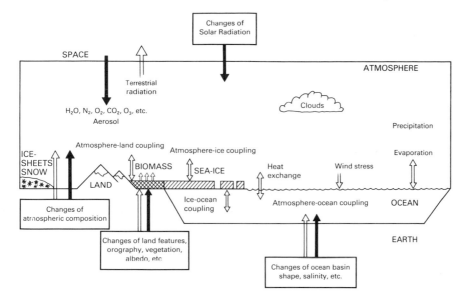

Figure 1.1 Schematic illustration of the components of the coupled atmosphere–ocean–ice–land climatic system. The full arrows are examples of external processes, and the open arrows are examples of internal processes in climatic change (*from GARP Joint Organizing Committee, 1975*).

tion of the climatic effects of changes in the sun's radiation, that is changes in the so called solar constant. There is now evidence that the solar constant varies very slightly over a period of a few years, and in particular with the number of sunspots. The solar constant has also probably varied over much longer time intervals.

On time-scales of tens of millions of years there are changes in the shapes of the ocean basins and the distribution of the continents as a result of sea-floor spreading and continental drift. Over geological time, these processes have resulted in substantial changes in global climate. On a shorter time-scale are the local modifications produced by man building cities and clearing forests.

Feedback mechanisms are very important in explaining short-term climatic changes. Such mechanisms act as internal controls of the climatic system, and display a coupling or mutual compensation among two or more elements of the climatic system. Feedback mechanisms may act either to amplify the value of one of the interacting elements (positive feedback) or to damp it (negative feedback). The best known positive feedback mechanism is the snow cover–albedo–temperature system, in which an increase of snow cover increases the surface albedo and thereby lowers the surface temperature. This decrease in surface temperature leads, all else being equal, to further increases in the extent of snow cover and still lower surface temperatures. Obviously, any positive feedback must be checked at some level by the intervention of other adjustment processes, or the climate would exhibit a runaway behaviour. The best example of a semi-permanent positive feedback process is the so-called greenhouse effect, in which the absorption of long-wave radiation by water vapour produces a higher surface temperature. Considerations of feedback mechanisms lead to the definition of climate as the equilibrium state reached by the elements of the atmosphere, hydrosphere, and cryosphere under a set of given and fixed input conditions. It will be suggested later in the book that there may be several possible equilibrium states for one given set of input conditions.

While the natural variations of climate have been larger than those that may have been induced by human activities during the past century, the rapidity with which human impacts threaten to grow in the future is a matter of concern, and is discussed in detail in Chapters 6 and 7. In particular the relative roles of changing atmospheric carbon dioxide and particle loading as factors in climatic change are discussed in the last two chapters. The amount of carbon dioxide in the atmosphere is steadily increasing because of the burning of fossil fuels. The present-day carbon dioxide excess relative to the year 1850 is about 13 per cent. Projects suggest that the CO_2 excess may reach around 30 per cent by 2,000 AD. Since carbon dioxide absorbs long-wave radiation and therefore exerts a greenhouse effect, increasing amounts must cause a global warming. Man is also increasing the total atmospheric loading of small particles, but their climatic effect is uncertain since they can cause either cooling or warming depending on the nature of the underlying surface. There are other possible impacts of human activities that should be considered in projecting future climates. One of these is the thermal pollution resulting from man's increasing use of energy and the inevitable discharge of waste heat into either the atmosphere or the ocean. Also widespread changes of surface land character resulting from agricultural use and urbanization may also have significant impacts on future climates.

References

GARP JOINT ORGANIZING COMMITTEE 1975: *The Physical basis of climate and climatic modelling.* GARP Publication Series **16**. Geneva: World Meteorological Organization.

LOCKWOOD, J. G. 1976: *The physical geography of the tropics: an introduction.* Kuala Lumpur: Oxford University Press.

US NATIONAL ACADEMY OF SCIENCES 1975: *Understanding climatic change, a program for action.* Washington, DC.

2
Radiation: The Prime Energy Source

1 General introduction

Energy may be defined formally as the capacity for doing work, and it may exist in a variety of forms including heat, radiation, potential energy, kinetic energy, chemical energy, and electric and magnetic energies. It is a property of matter capable of being transferred from one place to another, of changing the environment and is itself susceptible to change from one form to another. If nuclear reactions are excluded, it can be stated that energy is neither created nor destroyed, and from this it follows that all forms of energy are exactly convertible to all other forms of energy, though not all transformations are equally likely. It is therefore possible for any particular system to produce an exact energy account in which the energy gained exactly equals the energy lost, plus any change in storage of energy in the system.

Continual transformations of energy from one form to another take place in the atmosphere and on the earth's surface. These energy transformations are largely responsible for creating the natural environment because without energy the world would be completely dead, for there would be no movement or life. Indirectly, the sun drives all the meteorological and climatological wind systems, and these continually dissipate energy mainly by friction at the ground surface. It is estimated that, in the absence of solar radiation which creates atmospheric motions, dissipation of atmospheric movement by friction would be almost complete after six days.

2 Radiation laws

Any object not at a temperature of absolute zero ($-273\,°C$) transmits energy to its surroundings by radiation, that is, by energy in the form of electromagnetic waves travelling with the speed of light and requiring no intervening medium. This radiation is characterized by its wavelength, of which there is a wide range or spectrum extending from the very short X-rays, through the ultraviolet and visible to infrared, microwaves and radio-waves. The wavelengths of visible light are in the range $0.4\,\mu m$ to $0.74\,\mu m$ ($1\,\mu m = 10^{-4}\,cm$).

A valuable theoretical concept in radiation studies is that of the blackbody, which is one that absorbs all the radiation falling on it and which emits, at any temperature, the maximum amount of radiant energy. For a perfect all-wave blackbody, the intensity of radiation emitted and the wavelength distribution depend only on the absolute temperature, and in this case a number of simple laws apply.

2.1 Kirchhoff's law

This law states that the ratio of the emitted energy to the absorbed energy is solely a function of temperature T and the wavelength λ. Thus

$$\frac{E_{em}}{E_{ab}} = f(\lambda, T)$$

where E_{em} is the amount of energy emitted per unit area of the body, and E_{ab} is the fraction of incident energy which is absorbed by the body, i.e. neither reflected nor transmitted. When $E_{ab} = 1$, the body is referred to as a perfect blackbody, which absorbs all the energy incident upon it. It also follows that a blackbody possesses the highest emissivity at all wavelengths as compared with all other bodies.

2.2 Stefan–Boltzmann law

This law states that the total energy emission E by a unit area of blackbody surface in terms of its absolute temperature is:

$$E = \sigma T^4$$

Here σ is the Stefan–Boltzmann constant, which is $5{\cdot}6697 \times 10^{-12}$ W cm^{-2} K^{-4}. A body emitting radiation whose intensity is lower than that of a blackbody, by the same factor at all wavelengths, is referred to as 'grey'. The rate of total energy emission for a unit area of a greybody surface is given by:

$$E = \varepsilon \sigma T^4$$

where ε (always less than unity) is the emissivity of the greybody surface.

2.3 Wien displacement law

As shown in Figure 2.1, a blackbody does not radiate the same amount of energy at all wavelengths for any given temperature. At a given temperature, the energy radiated reaches a maximum at some particular wavelength and then decreases for longer or shorter wavelengths. The Wien displacement law states that this wavelength of maximum energy (λ_{max}) is inversely proportional to the absolute temperature, i.e.

$$\lambda_{max} T = \alpha$$

where α is a constant with a value of $0{\cdot}2897$ cm K if λ is in centimetres.

2.4 Solar and terrestrial radiation

The two types of radiation which are found in the atmosphere are considered in detail in the next sections. They are solar radiation from the sun, and terrestrial radiation from earth and atmosphere. Solar radiation is the source

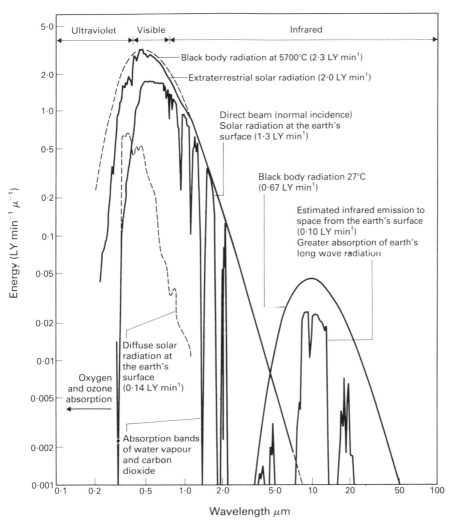

Figure 2.1 Electromagnetic spectra of solar and terrestrial radiation. The black-body radiation at 5,700 °C is reduced by the square of the ratio of the sun's radius to the average distance between the sun and the earth in order to give the flux that would be incident on the top of the atmosphere (*after Sellers, 1965*).

of energy which drives the atmosphere, whilst terrestrial radiation is its ultimate sink. Solar radiation is absorbed and scattered by the atmosphere, the oceans and the land surfaces. Terrestrial radiation is emitted by the atmosphere, by clouds and by the surface and is eventually radiated to space.

3 Short-wave radiation

It is usual to divide the entire atmospheric radiation regime into two parts—

the solar (or short-wave) regime and the terrestrial (or long-wave) regime. This is possible because of the greatly differing temperatures of the earth and sun. The high temperature of the sun (about 6,000 K) results in over 99 per cent of the solar energy being at wavelengths of less than 4 μm, whereas the much lower temperatures of the atmosphere and surface materials (generally < 300 K) yields most energy in the 4–100 μm region. Thus a division of the spectrum at about 4 μm effectively separates the two.

3.1 The solar source

Energy is generated in the sun by hydrogen being converted into helium by a complex series of nuclear reactions. Until fairly recently most writers would comment that the actual luminosity of the sun has appeared to be remarkably constant during the present century, and that the astrophysical theories of stellar structure and evolution suggest that such constancy should be expected. The amount of energy passing per unit time through a unit area at right angles to the direction of the solar beam outside the atmosphere must be known before it is possible to compute the energy emitted by the sun in all directions and in the direction of the earth. The concept of a 'solar constant' was introduced by A. Pouillet in 1837 in order to facilitate such calculations. The solar constant is defined as the quantity of solar energy at normal incidence outside the atmosphere at the mean sun–earth distance. The very name 'solar constant' suggests a unit which does not vary, but recent research has shown that small variations do occur. The correct value of the solar constant appears to be in the probable range 1360 $\pm$ 20 W m^{-2} (about 1·9 cal cm^{-2} min^{-1}).

The solar photosphere (Figure 2.2) is the 'surface' of the sun, as observed in white light, and its effective temperature is about 5,800 K. Above the photosphere is found the chromosphere, in which temperature increases outward, and beyond the chromosphere is the corona which stretches to great distances from the sun. Below the photosphere is the convective zone, in which energy from below is partly transported by convective motions, and probably reaches to depths of about 200,000 km. At greater depths, radiative transport of energy is again sufficient, and the plasma is in radiation equilibrium down to the convective core of the sun, where the thermonuclear processes, responsible for the sun's energy production, take place.

There are many types of solar activity including the emission of particles, X-rays and radio-waves over and above the level of the 'normal' quiet, undisturbed sun. Perhaps the best known are the recurrence of sunspots, flares and the changing aspects of the corona.

3.1.1 *Sunspots*

Most of the energy emitted by the sun comes from the photosphere. The main evidence of variation in this region is the occurrence of sunspots, which are darkened portions of the solar surface. Sunspots look dark because the temperature is a few thousand degrees Kelvin cooler than the surrounding photosphere. They may last for many days and are characterized by strong magnetic fields. Locally some radiation is blocked by sunspots, but this may

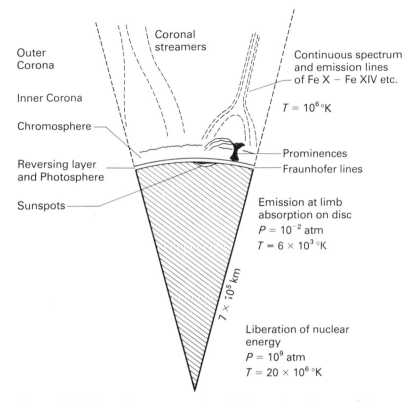

Outer Corona

Inner Corona

Chromosphere

Reversing layer and Photosphere

Sunspots

Coronal streamers

Continuous spectrum and emission lines of Fe X — Fe XIV etc.

$T = 10^6 \,°\text{K}$

Prominences
Fraunhofer lines

Emission at limb absorption on disc
$P = 10^{-2}$ atm
$T = 6 \times 10^3 \,°\text{K}$

7×10^5 km

Liberation of nuclear energy
$P = 10^9$ atm
$T = 20 \times 10^6 \,°\text{K}$

Figure 2.2 Schematic representation of the various layers of the sun (*after Robinson, 1966 and Ellison, 1956*).

not change the total solar luminosity. Because of the relatively short time required for diffusion throughout the convective envelope, any energy coming into the base of the envelope will be distributed over the entire surface of the sun. Thus, the blocking effect associated with variations in the number of sunspots is not likely to change the solar luminosity, because if the energy is not emitted at one place it will diffuse to some other place.

It has been known since the middle of the last century that the number of sunspots visible in the photosphere varies with time. During certain periods the sun is quiet and hardly any activity can be seen. These sunspot minima last for a few years, after which the activity picks up, and the number of sunspots present at any one time increases to a maximum, roughly 3–6 years after the minimum. During the next 6–7 years after a maximum, which may itself last a couple of years, the activity decreases, and a new minimum is reached. The whole period from one minimum to the next lasts roughly 11 years and is commonly called the 11-year cycle, even though periods as short as 9 years and as long as 14 years are observed. Sunspots change magnetic polarity with the coming of a new cycle, and since the magnetic field is of fundamental importance for an understanding of solar activity, there can also be considered to be a 22-year cycle. There is also evidence for even longer sunspot cycles. For instance, very high and very low peaks in the 11-year period occur with intervals of seven or eight 11-year cycles, and this secular variability is known as the 80-year period.

3.1.2 *Short-term variations in the solar constant*
There is marked disagreement among research workers as to whether or not there are very short-term variations in the value of the solar constant. Certainly any change in the efficiency of the convective energy transport in the convective envelope could alter the surface luminosity. Changes in the solar magnetic fields could well modify convection rates. At the time of writing no long-term record of the total solar radiation reaching the earth's orbit has been compiled from an extraterrestrial platform, so all the evidence is of an indirect nature. Direct measurements from the surface of the earth are difficult because of the presence of the atmosphere which is not completely transparent.

One of the longest sets of data available on solar luminosity consists of the calorimetric measurements made over a period of 30 years by Abbot *et al.* (1942) at the Smithsonian Astrophysical Observatory. They were attempting to detect variations in the solar luminosity and felt that their measurements were indicative of time variations with an amplitude of 1 to 2 per cent. The bulk of the amplitude of this variance is due to a large drop during the middle 1920's, while the remaining amplitude is about $\frac{1}{2}$ per cent. It is not clear what caused the 1924 drop in the measured solar constant, since it occurred simultaneously in the space of a few months in both the northern and southern hemispheres.

An indirect method of monitoring the solar luminosity is by measuring the average brightness of uranus and neptune. Some of the difficulties in measuring the solar constant arise because the sun is bright and its radiation is emitted from an extended disc which must be entirely included in the detecting aperture. Another problem is the lack of a convenient reference with which the sun can be compared. The outer planets, uranus and neptune, have small apparent diameters and are faint enough to be compared conveniently to stars. Brightness variations of 1 per cent have been reported by Lockwood (1975) for these planets. These variations are correlated with each other and represent either the effect of variable solar luminosity or some effect of the solar wind.

Baur (1964) has reported that the monthly solar constant values reported by Abbot appear to show a systematic variation within the 11-year sunspot cycle over a range of about $\frac{1}{2}$ per cent. A similar result has been found by Kondratyev and Nikolsky (1970) using balloon observations of the solar constant. They found that the solar constant is more than 2 per cent lower for no sunspot activity than for moderate activity and decreases again for high levels of activity. Their results can be summarized in the relationship:

$$S(N) = 1 \cdot 903 + 0 \cdot 011 N^{\frac{1}{2}} - 0 \cdot 0006 N \text{ cal cm}^{-2} \text{ min}^{-1}$$
$$(1 \text{ cal cm}^{-2} \text{ min}^{-1} = 698 \text{ W m}^{-2})$$

where $S(N)$ is the value of the solar constant corresponding to a Wolf number N. The Wolf relative sunspot number N is defined by

$$N = k(10g + f)$$

where g is the number of groups of sunspots, f is the total number of sunspots present and k is a constant.

Mean annual sunspot (Wolf) numbers have been recorded for many years by astronomers and are given in Figure 2.3. The remarkable feature of Figure

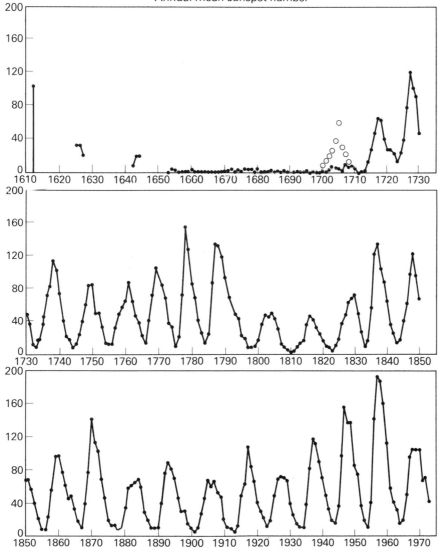

Figure 2.3 Mean annual sunspot (Wolf) numbers (*after Schneider and Mass, 1975*).

2.3 is the relative absence of sunspots from about 1650 to about 1700, the so called Maunder minimum. Eddy (1976) who has delved into rare book collections to uncover accounts of astronomers at that time, concludes that the Maunder minimum may well be an indication of the absence of sunspots, not simply a dearth of measurements.

Other evidence for short-term variations in solar luminosity comes from the field of radiocarbon dating using ^{14}C. Radioactive ^{14}C atoms continually originate in the high atmosphere by neutrons colliding with atoms of ordinary atmospheric nitrogen, ^{14}N, some of which lose a proton and are thus converted into radioactive carbon. The neutrons responsible are themselves produced by cosmic ray particles from outer space bombarding the outer atmosphere. The incidence of cosmic rays is reduced at times of solar disturbance, probably through their being deflected away from the earth by powerful magnetic fields. Repeated radiocarbon measurements on objects of known ages have shown those originating between about AD 1300 and AD 1800 to be too rich in ^{14}C by 1–3 per cent, apparently implying a quiescence of solar flare activity, and a relationship with the cold period between about 1430 and 1850.

3.1.3 *Long-term variations in the solar constant*

Energy is generated in the sun by hydrogen being converted into helium by a complex series of nuclear reactions, and following many of these reactions neutrinos are emitted which readily escape from the interior of the sun and may be detected on earth. The best-evolved models of the sun recently available predicted a greater number of neutrinos than those which were actually observed in experiments on earth. Revised models of the sun which take into account the low neutrino-count assume that variations take place in the internal structure of the sun which result in both changes in the number of neutrinos and also in the solar luminosity. Such theories tend to assume that the solar luminosity decreases for a few million years and then returns to normal. An example by Ezer and Cameron (1972) is shown in Figure 2.4. In their work, a model of the sun that had been evolved for $4{\cdot}6 \times 10^9$ years had the central 0·56 of its mass suddenly mixed. This was considered to be an atypical mixing, because it brings large amounts of fresh hydrogen toward the centre of the sun, which allows the centre of the sun to expand and relax toward a new equilibrium state. After this relaxation had occurred, Ezer and Cameron then mixed the central region of the sun again, producing a result that should be more typical of periodic mixing in the interior of the sun. This reduced both the neutrino flux and the solar luminosity. According to models of this type the sun is at present near a minimum of luminosity and would normally be much more luminous. Thus changes in solar luminosity, corresponding to changes in solar structure, probably explain long-term changes in global climate, and particularly the very warm climate in the Cretaceous and early Tertiary when the poles were largely ice-free, and the present rather glacial climate.

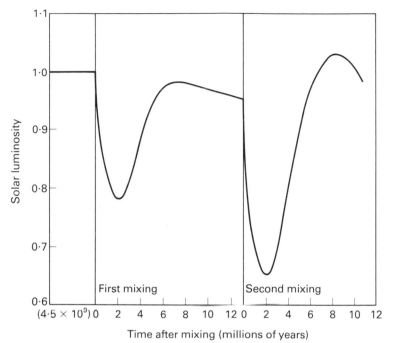

Figure 2.4 Variations of the solar luminosity during the numerical experiments carried out by Ezer and Cameron, 1972.

3.2 The Milankovitch mechanism

The present orbit of the earth is slightly elliptical with the sun at one focus of the ellipse, and as a consequence the strength of the solar beam reaching the earth varies about its mean value. At present the earth is nearest to the sun on 2–3 January and farthest from the sun on 5–6 July. This makes the solar beam near the earth about $3\frac{1}{2}$ per cent stronger than the average solar constant in January, and $3\frac{1}{2}$ per cent weaker than average in July. Now the gravity of the sun, the moon and the other planets causes the earth to vary, over many thousands of years, its orbit around the sun. Three different cycles are present, and when combined, produce the rather complex changes observed.

Firstly the earth's orbit varies from almost a complete circle to a marked ellipse, when it will be nearer to the sun at one particular season. A complete cycle from near-circular through a marked ellipse back to near-circular takes between 90,000 to 100,000 years. When the orbit is at its most elliptical, the intensity of the solar beam reaching the earth must undergo a seasonal range of about 30 per cent.

Secondly the earth's axis wobbles so that the season of the closest approach to the sun varies. The complete cycle takes about 21,000 years, so 10,000 years ago the northern hemisphere was in summer, rather than winter as at present, when the earth was closest to the sun.

Thirdly the tilt of the earth's axis of rotation relative to the plane of its orbit is believed to vary at least between 21·8° and 24·4° over a regular period of about 40,000 years. At present it is almost 23·44° and is decreasing by about 0·00013° a year. The greater the tilt of the earth's axis, the more pronounced is the difference between winter and summer.

The three mechanisms just described are sometimes known as the Milankovitch mechanism after Milankovitch (1930) who first described them in detail. The Milankovitch mechanism affects only the seasonal and geographical distribution of solar radiation on the earth's surface, yearly totals remaining constant. Surplus in one season is compensated by a deficit during the opposite one; surplus in one geographical area is compensated by simultaneous deficit in some other zone.

3.3 Attenuation of solar radiation in the atmosphere

The spectrum of energy received from the sun at the ground surface is found to depart from the ideal form for a blackbody at the sun's temperature. This is partly due to absorption by certain chemical elements in the cooler parts of the solar atmosphere, but it is largely due to absorption in the earth's atmosphere. It is therefore necessary to consider the attenuation of solar radiation in the atmosphere in some detail.

3.3.1 *The Lambert–Bouguer law of transmission*

As a beam of radiant energy propogates through the atmosphere from a source at x_1, to a receiver at x_2, there is a loss of energy because of scattering and absorption within the atmosphere. The fraction of monochromatic energy that is transmitted along a homogeneous path is given by the Lambert–Bouguer law:

$$T = \exp(-\beta x)$$

where T is the fraction of transmitted monochromatic energy, β is the volume extinction coefficient, and x is the path length $(x_1 - x_2)$. β, at sea-level, is of the order of 0·1 km^{-1}, in which case 37 per cent of the energy is directly transmitted in a horizontal distance of 10 km.

It is convenient to replace the product βx by one parameter (τ') called the optical thickness, i.e. $\tau' = \beta x$. Since, in general, β is not constant along the path, the optical thickness is calculated from the integral sum of the various values. T is then given by:

$$T = \exp(-\tau') = I/I_o$$

where I_o and I are the initial and final intensities of the radiation.

When using this relationship for the attenuation of solar radiation in the atmosphere it is usual to allow for the path length of the beam through the atmosphere. This path length will be at a minimum when the sun is overhead and will increase as the sun sinks towards the horizon. Thus even with a constant value of τ' per unit path length of atmosphere, the attenuation of the solar beam will vary with the angle of the sun from the zenith. This may be corrected for by introducing the optical air mass m. The optical air mass is equal to 1 when the sun is overhead and increases with increasing solar angle. When the zenith angle is 60° it equals 2, indicating that the path length through the atmosphere is twice that when the sun is overhead. To a good approximation

$$m = \sec z$$

where z is the angle between the local zenith and the direction of the sun. Thus in the atmosphere:

$$I = I_o \exp(-\tau m)$$

when τ is the optical thickness of the atmosphere measured in the local zenith direction.

Another commonly employed form of the above equation is:

$$I = I_o q^m$$

when q is the transmissivity or the transmission factor of the atmosphere.

It is often useful to separate τ into its various components. Atmospheric attenuation is due to both scattering (τ_{sc}) and absorption (τ_{ab}) of the incident ray, thus:

$$\tau = \tau_{sc} + \tau_{ab}$$

Furthermore, the total optical thickness equals the sum of the separate optical thicknesses of the attenuating constituents:

$$\tau = \tau_G + \tau_{O_3} + \tau_{H_2O} + \tau_A$$

where τ_G is the optical thickness of the permanent atmospheric gases such as argon, nitrogen, oxygen, etc., τ_{O_3} is the optical thickness of ozone, τ_{H_2O} is the optical thickness of water vapour, and τ_A is the optical thickness of atmospheric aerosols.

The optical thickness of a clear dry atmosphere will be equal to τ_G. The addition of dust and water vapour increases the optical thickness and therefore the attenuation. One useful measure of this increase in attenuation due to dust, water vapour etc., is the turbidity factor (τ/τ_G).

The turbidity of the atmosphere over the British Isles has been investigated by Unsworth and Monteith (1972) using a turbidity coefficient τ_a defined as follows:

$$\tau_a = -\frac{1}{m}\ln\left(\frac{I}{I^*}\right)$$

where I is the observed direct solar radiation normal to the solar beam, and I^* is the calculated theoretical direct radiation below a clean atmosphere containing known amounts of absorbing and scattering gases. Unsworth and Monteith calculated I^* from a knowledge of the following:

(i)   solar constant
(ii) sun–earth distance
(iii) total water vapour, carbon dioxide and ozone contents of the atmosphere
(iv) the optical air mass which is related to solar altitude.

Theoretical values of $I*$ may be obtained from Figure 2.5, parts (a) and (b). The observed values of I will fall substantially below the clean air values calculated for $I*$ and may be obtained from part (c) of Figure 2.5. According to Unsworth and Monteith, τ_a at remote coastal sites in the UK, probably ranges from about 0·05 in clear polar air streams to about 0·35 in air streams of continental origin. At rural sites inland, τ_a is probably about 0·05 higher on all occasions, because of natural aerosol originating nearby, and in polluted urban areas may be even higher.

3.3.2 *Scattering*

The simplest atmospheric model is that of a non-absorbing medium in which the scattering particles are all of a size much smaller than the wavelength of the incident light, a criterion which applies principally to molecules of the atmospheric gases. Under these conditions the Rayleigh scattering theory is appropriate, in which the maximum scattering occurs in both the forward and backward directions relative to the incident beam. Also the efficiency of scattering in a clear atmosphere is critically dependent on the wavelength of the light, in the sense that short wavelenghts are scattered much more strongly than are long wavelengths. This fact, in combination with the spectral sensitivity of the eye and the spectral distribution of sunlight, accounts for the blue colour of a clear day-time sky.

When the scattering particles are small in comparison with the wavelength of the light, the Rayleigh scattering theory can be applied. However, in the range of particle radius varying from a tenth of a wavelength through to about 25 wavelengths it is necessary to use the theory developed by Mie. In Mie's theory the amount of light scattered in the forward direction of the beam is much larger than that scattered in the backward direction. Also, in the presence of large particles, light of all colours is scattered to a greater extent, so that the sky will appear less blue and eventually will become white when a sufficient number of large particles are present.

3.3.3 *Global solar radiation*

The rate at which radiation falls on a unit area of a plane surface is known as the 'irradiance'. The irradiance at any point in the atmosphere will vary with the orientation of the receiving surface and this orientation must therefore always be specified or clearly implied. Global solar radiation is defined as the solar irradiance on a plane horizontal surface coming from the whole hemisphere. Global solar radiation is made up of two components: firstly a component coming directly from the sun and secondly a diffuse component from the remainder of the hemisphere. If there is cloud between the observation point and the sun then the direct component is zero and there is only a diffuse component.

Direct solar radiation is the solar irradiance on a surface perpendicular to the direction of the sun and continued in a narrow solid angle centred on the sun's direction. The actual amount of direct radiation depends on the cloudiness and also on the solar altitude. In contrast, diffuse radiation may be considered as coming equally from all parts of the sky. Global radiation (R_G),

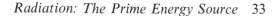

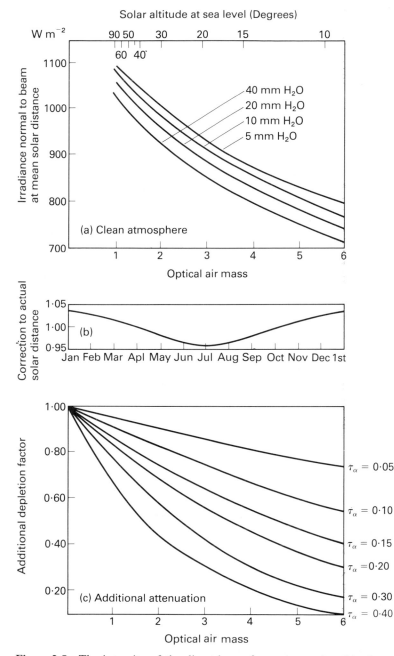

Figure 2.5 The intensity of the direct beam for a given solar altitude, precipitable water content, time of year, and turbidity coefficient can be found by multiplying the appropriate values from graph (a) by the correction factors in graph (b) and graph (c) (*after UK Section International Solar Energy Society, 1976*).

direct radiation (R_{DIR}) and diffuse radiation (R_{DIFF}) are related by the equation:

$$R_G = R_{DIR} \sin \theta + R_{DIFF}$$

where θ is the solar altitude.

3.4 Global pollution and solar radiation

Variations in the flux of incoming solar radiation under cloudless skies at the surface of the earth may be either due to natural causes or induced by man's activities. Volcanic eruptions are the main natural cause of atmospheric pollution while man influences solar radiation by burning fossil fuels, causing industrial pollution and by dust storms resulting from dry land farming. Hence the turbidity of the atmosphere measured at any particular site depends partly on the local weather and partly on the synoptic history of the prevailing air mass. The local weather determines the input of aerosols from domestic and industrial sources and the synoptic history the input of aerosols and water vapour from much more distant sources and their distribution throughout the atmosphere. Major changes in turbidity at rural sites are usually the result of changes in air mass type, and local sources of aerosol, either natural or man-made, are seldom responsible for comparable changes in the attenuation of solar radiation. Much of the change in turbidity between air masses is due to changes in water vapour content. Thus moist maritime tropical air has a considerably greater turbidity than cold arctic air. Similarly, turbidity tends to be lower in winter than in summer because of the greater absolute atmospheric water content in summer. The influence of water vapour content can be seen clearly in Figure 2.6 which shows observations from Mauna Loa Observatory, Hawaii, a location remote from known sources of pollution (Pueschel *et al.* 1974). Direct radiation measurements were made when the sky was completely cloud-free in the direction of the sun which was at a zenith angle of 60° (corresponding to an air mass of 2), while atmospheric precipitable water measurements were based on observations in a water vapour absorption band. The most immediate and striking conclusion from Figure 2.6 is the nearly perfect (inverse) tracking between solar radiation and water vapour content of the atmosphere on a day-to-day basis. Excluding the dates indicated, the correlation coefficient between the two parameters is 0·95. The dates marked on Figure 2.6 are periods when direct solar radiation was attenuated to lower values than would be expected from a consideration of only the precipitable water absorption. One period according to Pueschel *et al.* was due to blowing snow at the station; the rest were due to volcanic haze from the nearby Kilauea Volcano.

This particular problem is discussed further in Chapter 6.

3.4.1 *Volcanos and solar radiation*

Recent research (Russell *et al.* 1976) has led to a picture of a broad maximum in stratospheric aerosol concentration centred at an altitude of approximately 20 km with occasional appearances of higher layers. The single most abundant constituent of the particles is sulphate, existing primarily as sulphuric acid,

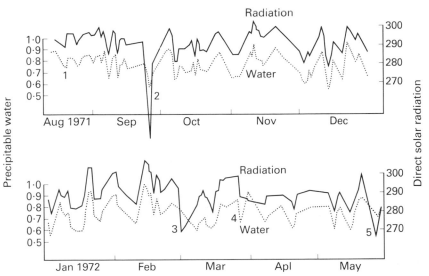

Figure 2.6 Day-to-day variations at Mauna Loa Observatory of direct solar radiation and precipitable water ratio at sun's altitude of 30° (air mass of 2). In all cases the sky was cloud-free in the direction of the sun. Total atmosphere precipitable water based upon the relative intensities of the sun in a water vapour band (0·935 μm) and a nearby region (0·880 μm), therefore the higher the value the drier the atmosphere (*after Pueschel et al., 1974*).

although other materials are frequently present as trace constituents. A significant fraction of the particles form locally from chemical reactions of sulphur-bearing gases such as sulphur dioxide and hydrogen sulphide.

Concentrations of stratospheric trace constituents including both fine particles and gases, are often greatly enhanced by violent volcanic eruptions, and the perturbations may last for several years (Cadle *et al.* 1976). The particulate matter from such eruptions is largely volcanic ash and sulphate, especially impure sulphuric acid droplets. The gases are largely water vapour, carbon dioxide, sulphur dioxide, carbon monoxide, hydrogen, and halides, especially hydrogen chloride. Much of the sulphur dioxide is oxidized and hydrated to form additional sulphuric acid. The particulate material can affect the radiation balance of the earth.

Volcanic eruptions are of the two main types: effusive eruptions and explosive eruptions. Effusive eruptions are those in which lava flow predominates. Explosive eruptions do not necessarily produce any flow of liquid lava, since it may all be thrown vertically upwards to condense as spray, and solidify as ash and cinders, and the partly solid froth to solidify and fall as pumice. It is the explosive eruptions which are of most meteorological interest since they can have a world-wide influence. The effects of lava flows are local and are of little meteorological interest.

Explosive eruptions can throw vast quantities of material to great heights. For example in May 1970, Hekla Volcano, Iceland, erupted and within an hour of the start of the eruption, dust clouds monitored by radar from the naval air base at Keflavik had reached an elevation of 16 km. Similarly, US Air Force aircraft and ground-based radar reported that Mount Redoult, Alaska, ejected dust clouds to heights of 12 to 13·7 km during an eruption between 25 January and 9 February 1966. The actual volume of material blown into the atmosphere during an explosive eruption can amount to 1 km³ or more. After the explosive eruption of Krakatoa in 1883, complete darkness existed for a period at noon at Bandung, Java, 250 km away, a couple of hours after the greatest explosion. During the eruption of Hekla, Iceland, in 1947 dust fell on ships in the Atlantic 800–1,000 km away to the south and east, and three days later in Finland.

The residence time in the atmosphere of the dust particles depends on their size and altitude. Very large objects such as stones will fall out very quickly as will other large particles, so that eventually only fine dust will remain. Even fine dust in the troposphere is very quickly washed out by rain and snow, and therefore it does not accumulate but remains only for a few days. In contrast, fine dust in the stratosphere, where there are no clouds, only slowly settles out by gravity and can have a residence time of several years. Radii of solid particles observed in persistent stratospheric dust veils are mainly of the order of 0·5 μm, which is about the same magnitude as the wavelengths of the bulk of the solar radiant energy.

Volcanic dust and gases ejected into the stratosphere are swept towards the east by the powerful zonal (east–west) winds. Typically, volcanic dust takes from 2 to 6 weeks to circuit the earth in middle or lower latitudes, and from 1–4 months to become a fairly uniform veil over the whole of the latitude zone swept by the wind system into which it is injected. Dust in the lower equatorial stratosphere is gradually spread over the whole earth by the mean meridional circulation of the atmosphere. Volcanic dust originating in high latitudes hardly spreads in significant quantities beyond about latitude 30° in the hemisphere of origin; though small quantities may reach any part of the earth. The final stage of a world-wide dust veil from an equatorial eruption is probably a concentration of the last remaining airborne dust over the two polar caps.

Because of their relatively large size compared with the incident shortwave radiations, volcanic dust particles predominantly scatter solar radiation in the direction of the incident beam. This is the so called Mie effect discussed earlier. The result is that the direct radiation is reduced much more substantially by scattering than is the global solar radiation, since the bulk of the radiation scattered from the direct beam re-appears at the surface as diffuse radiation. This is shown in Figure 2.7 which illustrates the changes which took place in the direct, diffuse and global radiation at Aspendale, Melbourne (38° S, 145° E), after the eruption of Mount Agung (Bali) in 1963. The Bali 1963 dust veil was the most effective since 1902–3 and possibly since 1883–6. A marked drop in the direct beam radiation is largely compensated by a similar large increase in the diffuse components. The changes in the direct and diffuse components almost completely cancel out and only a very slight fall is detectable in the global radiation curve. Budyko (1974) quotes the values given in Table 2.1 for the ratio of the depletion of the global radiation received at the

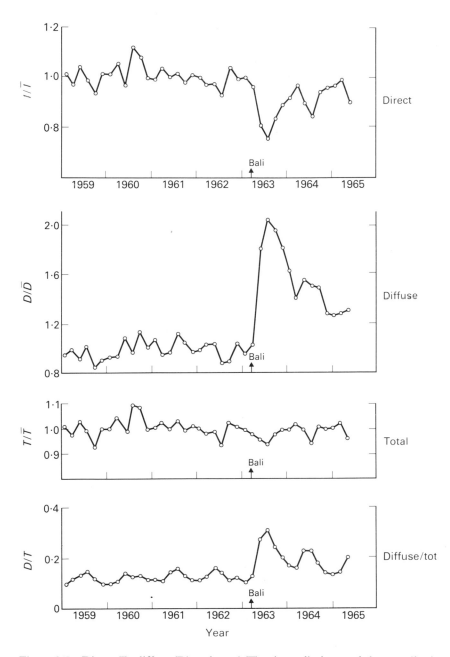

Figure 2.7 Direct (I), diffuse (D) and total (T) solar radiation, and the contribution of the diffuse (sky) radiation to the total at Aspendale, Melbourne (38° S, 145° E) after the Bali 1963 eruption. Fractions of the 1959–62 averages (*after Dyer and Hicks, 1965*).

Table 2.1   Effect of dust on the radiation regime (*after Budyko, 1974*).

Latitude (degrees)	Decline of total radiation / Decline of direct-beam radiation
90	0·24
80	0·23
70	0·22
60	0·21
50	0·19
40	0·18
30	0·16
20	0·14
10	0·13
0	0·13

surface to the depletion of the direct solar beam at various latitudes, in the presence of a layer of silica dust in the stratosphere. The dust particles are assumed to have diameters between 0·02 and 0·3 μm. It is seen that the decline in global radiation is small and also that it is greatest at high latitudes.

Pueschel *et al.* (1974) have compared the influence of local and distant volcanic eruptions on solar radiation. They compared the local effects of eruptions from Kilauea, Hawaii, on the solar radiation record at Mauna Loa Observatory, with the global effects of the eruption of Mount Agung on the Island of Bali in 1963. Typical Mauna Loa values show a 5–15 per cent decrease of direct solar radiation, lasting for a few days, immediately after new eruptions of Kilauea. In contrast, after the eruption of Agung, there was a 25 per cent decrease in the monthly average direct radiation and a 100 per cent increase in the diffuse sky radiation at Aspendale (Melbourne), lasting over several years. A similar increase in turbidity of smaller magnitude apparently due to the intrusion of Agung's dust into the stratosphere has been found by Ellis and Pueschel (1971) from an analysis of the monthly averages of Mauna Loa's solar radiation measurements. The conclusion is that on the natural scale volcanic eruptions of the calibre of Krakatoa (1883) or Agung (1963) are required to cause atmospheric dust loadings that would affect solar radiation in excess of the day-to-day variations due to water vapour and local volcanic effluents.

3.4.2 *Lamb's dust veil index*

According to Lamb (1970), the available types of observation of the effects of volcanic explosions afford three alternative ways of arriving at a numerical assessment of dust veil magnitude. These have been expressed by Lamb using the following formulae for a dust veil index (DVI) in which the coefficients are adjusted to give a DVI of 1,000 for the 1883 eruption of Krakatoa (6° S, $105\frac{1}{2}$° E). The most reliable assessments are obtained by using as many of these formulae as can be applied to a given eruption and taking a round figure for the arithmetic mean as the index of order of magnitude:

$$\text{DVI} = 0{\cdot}97 \cdot R_{d\,max} \cdot E_{max} \cdot t$$

$$\text{DVI} = 52 \cdot 5 . T_{d\,max} . E_{max} . t$$

$$\text{DVI} = 4 \cdot 4 . q . E_{max} . t$$

where $R_{d\,max}$ is the greatest percentage depletion of the direct solar beam by any monthly average for the middle latitudes of the hemisphere concerned after the eruption. $T_{d\,max}$ is the estimated lowering of average temperature for the most affected following year in the middle latitudes of the hemisphere concerned. q is the estimated volume in km^3 of solid matter dispersed as dust in the atmosphere. E_{max} is the greatest proportion of the earth at some time covered by the dust veil (taken as 1 for eruptions between 20° N and 20° S, 0·7 for eruptions between latitudes 20° and 35°, 0·5 for latitudes 35°–42°, 0·3 for latitudes greater than 42°). t is the total time in months elapsed after the eruption to the last readily observed effects in middle latitudes (optical effects cease and temperature and radiation measurements return to their former level).

Lamb comments that no more refined expressions, using exponential decay rates and the like, would be appropriate to the quality of the data available for most eruptions in the past.

4 Long-wave radiation

Both the ground and the atmosphere, as any other bodies above absolute zero, radiate energy. Compared with the sun, both the earth's surface and the atmosphere are relatively cool, and their emissions lie in the invisible infra-red region of the spectrum. The average temperature of the earth's surface is about 285 K, and therefore most of the radiation is emitted in the spectral range from 4 to 50 μm, with a peak near 10 μm, as indicated by the Wien displacement law. This radiation is sometimes called terrestrial radiation, because it is emitted by the earth and atmosphere, and sometimes thermal radiation, because it is a form of heat energy.

Neither the earth's surface nor the atmosphere can be regarded as black-bodies. The earth's surface is commonly assumed to emit and absorb in the infrared region as a greybody, that is, as a body for which the Stefan–Boltzmann law takes the form

$$F = \varepsilon \sigma T^4$$

where the constant of proportionality ε is defined as the infrared emissivity or, equivalently, the infrared absorptivity. Practically all surfaces, including snow, have emissivities of between 90 and 95 per cent.

The radiation from the terrestrial surface is considerably less than that from the sun's surface, but nevertheless is comparable to the solar radiation flux that reaches the earth's surface. Obviously it depends on temperature, but is of the order of 350–400 W m^{-2}.

Atmospheric infrared emissions are more complex than those from the surface. Radiation is emitted only by those gases which absorb it strongly, i.e. water vapour, carbon dioxide, ozone and oxygen. In addition, the radiation and absorption of each of these gases has a complicated selective character. In addition to the basic gases that absorb infrared radiation the atmosphere contains traces of a number of other gases which absorb strongly in the infrared. These include the various nitrogen oxides (NO, N_2O, N_2O_4, N_2O_5), some hydrocarbons (C_3H_8, C_2H_6, C_2H_4, CH_4) and other gases.

Water vapour has the widest and strongest absorption bands, and plays the main part in the absorption of infrared radiation. The strongest absorption bands occur at 5·5–8·0 μm and beyond 20 μm, while in the so-called water vapour window between 8 and 13 μm the atmosphere is practically transparent to infrared radiation. Carbon dioxide absorbs strongly between 14–16 μm. The ground surface, on the other hand, emits approximately as a blackbody at all wavelengths. Thus the long-wave radiation emitted to space from the ground and clear atmosphere is composed mainly of that emitted by the atmosphere in the wavelengths of the strong absorption bands plus that which is emitted by the surface and transmitted outwards in the regions of weak atmospheric absorption. The long-wave radiation which reaches the ground from a clear sky above is mainly due to the emission by atmospheric gases in the first 1,000 m of the atmosphere. The relatively dense liquid water clouds in the atmosphere act essentially as blackbodies, and thus strongly modify both the upward and downward radiation.

It is seen (Figure 2.8) therefore that the net outgoing long-wave radiation (incoming − outgoing) from the ground surface has two basic components: the total long-wave emission from the surface, which is a function of the surface emissivity and temperature; and the counter long-wave radiation from the atmosphere which is a function of air temperature, water vapour content,

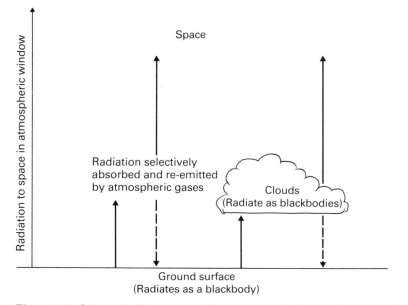

Figure 2.8   Schematic illustration of long-wave radiation exchanges in the atmosphere.

and the cloud cover. The counter radiation is almost of the same magnitude as the radiation from the ground surface, and only about 9 per cent of the infrared radiation from the ground surface escapes directly to space, mainly in the atmospheric window between 8 and 13 μm.

When considering the effect of cloud on atmospheric counter radiation there are two main components, which are the cloud amount and the temperature of the cloud base. Cloud layers are normally considered to radiate as blackbodies, and therefore the effect of cloudiness on the net long-wave radiation is often taken into account by multiplying the estimate for a clear sky by a factor $(1 - \lambda c)$ in which c is the fractional cloudiness and λ depends on cloud height. Cloud height is important since high-level clouds have much lower base temperatures than low-level clouds. For low clouds $\lambda = 0.8–0.9$, for medium clouds $0.6–0.7$ and for cirrus clouds about 0.2. Normally dense low clouds are most effective in retarding temperature falls at night, while cirrus clouds have little effect. Thus ground frost usually occurs on calm nights with little or no low cloud.

Under cloudless skies the main factor influencing the net long-wave radiation loss, apart from surface temperature, is the water vapour content of the lower atmosphere. Thus Brunt (1941) suggested that the atmospheric counter radiation (R_{LC}) could be estimated from

$$R_{LC} = \sigma T^4(0.52 + 0.065\sqrt{e})\ \text{W m}^{-2}$$

where T (K) and e (mbar) are the screen temperature and vapour pressure respectively, and σ is the Stefan–Boltzmann constant. The drier the lower atmosphere the greater will be the long-wave radiation loss. Since water vapour content tends to be correlated with temperature, it also follows that the long-wave loss will be greater with cool atmospheres than with warm from surfaces of equal temperature. This provides an interesting climatic feedback mechanism. The work of the CLIMAP project members (1976) and Gates (1976a, 1976b) suggests that for July, 18,000 BP during the last ice age, the water vapour content of the atmosphere and the rates of evaporation and precipitation were greatly reduced below their present-day values (Table 2.2).

Table 2.2 Summary of area-averaged simulated climatic variables for July at 18,000 BP and present time (*after Gates, 1976a, 1976b*).

Variable	Ice-age average		Difference (ice-age minus present) average	
	Northern hemisphere	Southern hemisphere	Northern hemisphere	Southern hemisphere
Surface air temperature (°C)	18·0	7·1	−8·3	−4·5
Cloudiness (%)	22·5	44·2	−2·9	−2·2
Precipitable water (mm)	14·2	12·9	−8·3	−3·9
Evaporation (mm day⁻¹)	4·0	3·5	−0·5	−0·9
Precipitation (mm day⁻¹)	4·5	3·1	−1·2	−0·1

A few simple calculations show that the global fall in average sea-surface temperature at 18,000 BP (about 2·3 degrees C) would probably decrease the infrared radiation from the ocean surface to the atmosphere by 2–3 per cent. Much of this infrared radiation is absorbed by the atmosphere and re-radiated, but with the lower water vapour contents at 18,000 BP the amount of absorption would probably fall by up to 10 per cent, allowing a marked increase in the direct cooling of the oceans by infrared radiation to space. Even with the present-day solar constant, the net radiation at the ocean surface at 18,000 BP would be less than at present, the rate of evaporation lowered, and the intensity of the hydrological cycle correspondingly reduced. The original fall in atmospheric water vapour content was due to the general fall in atmospheric temperatures, but it would in turn lead to a world-wide fall in temperature, thus intensifying the original change. This is an example of the variation in intensity of the greenhouse effect, which is discussed in detail in Section 6 of this chapter.

5 Interaction of radiation with surfaces

Radiation falling on a surface may be partly reflected, partly absorbed and partly transmitted. Most natural solid objects are opaque, so light is either reflected or absorbed. In contrast, water is translucent and light penetrates into the surface layers of the oceans, while the atmosphere is nearly transparent to short-wave radiation.

Radiation being received from a surface may have resulted from either reflection or radiation by the surface, or indeed both. It is essential to distinguish clearly between reflected and re-radiated radiation. If radiation is absorbed by the surface and then re-radiated, the wavelengths of the re-radiated radiation will vary according to the Stefan–Boltzmann and Wien laws, that is, it will be controlled by the absolute temperature of the surface and by its emissivity. Thus most natural objects radiate in the infrared. If radiation is directly reflected there is no change in wavelength, and so short-wave radiation is reflected as short-wave radiation. The ratio of the reflected to the incoming short-wave radiation is known as the albedo, and some typical values of albedo are shown in Table 2.3, where it is seen that the albedo

Table 2.3   Values of albedo for various surfaces (*after Budyko, 1974*).

Surface	Albedo (percentage of incoming short-wave radiation which is reflected)
Fresh, dry snow	80–95
Sea ice	30–40
Dry light sandy soils	35–45
Meadows	15–25
Dry steppe	20–30
Coniferous forest	10–15
Deciduous forest	15–20

of a blackbody is zero. The albedo of a vegetation canopy depends on its geometry and on the angle of the sun as well as on the radiation properties of its components. According to Monteith (1973), the albedo of relatively smooth vegetation surfaces, such as closely cut lawns, is around 25 per cent. For crops growing to heights of 50 to 100 cm, the albedo is usually between 18 and 25 per cent when the ground cover is complete, while the albedo drops to as low as 10 per cent over forests. The explanation for this decrease in albedo with increasing vegetation height is found in terms of the trapping of radiation by multiple reflection between adjacent leaves and stems. For the same reason, the albedo of most types of vegetation changes with the angle of the sun. Minimum values are recorded as the sun approaches its zenith and increases as the sun descends to the horizon because there is less opportunity for multiple scattering between elements of the canopy.

If there is no heat advection or conduction, the temperature of a surface will adjust itself so that as much energy is being radiated as is being absorbed, and thus the body will neither gain nor lose energy in the long term. The mean surface temperature of the earth is such that the earth radiates as much energy to space as it receives from the sun, so the long-term energy content of the planet is almost constant. In the last example it should be remembered that the intensity of solar radiation is greatly reduced with distance from the sun, and therefore the further planets are from the sun, the lower the surface temperature required to achieve radiation balance. Therefore planets near the sun, such as mercury (Table 2.4), are intensely hot, while distant planets, such as neptune, are extremely cold.

Table 2.4   Mean planetary temperatures.

Planet	Mean distance from sun (earth = 1)	Mean planetary albedo (percentage)	Mean temperature (K)
Mercury	0·387	6	616
Venus	0·723	76	235
Earth	1·000	30	254
Mars	1·524	16	209
Jupiter	5·20	73	105
Saturn	9·55	76	78
Uranus	19·2	93	55
Neptune	30·1	94	43
Pluto	39·4	14	42

Of particular interest to the meteorologist is the so called net radiation, which is the difference between the total incoming and the total outgoing radiation. The net radiation indicates whether net heating or cooling is taking place; the net radiation will normally be negative at night indicating cooling, but during the day it may be positive or negative depending on the balance of the incoming and the outgoing radiation.

In Figure 2.9 which illustrates the components of the net radiation balance, it is assumed that the soil is dry, that is, there is no evaporation nor condensation. The simplest case occurs at night, because there is no incoming short-

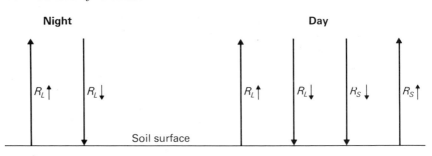

$R_L\uparrow$ Infrared radiation from surface

$R_L\downarrow$ Infrared counter radiation from atmosphere

$R_S\downarrow$ Incoming solar radiation

$R_S\uparrow$ Reflected solar radiation

Figure 2.9   Schematic illustration of the radiation balance of a soil surface.

wave radiation, but instead there is a continuous long-wave radiation loss. At night the soil surface emits long-wave radiation ($R_L\uparrow$) to the atmosphere, where water vapour and carbon dioxide absorb large amounts of this long-wave radiation, which in turn is partly re-radiated downwards ($R_L\downarrow$) to be re-absorbed by the soil surface. The difference between $R_L\uparrow$ and $R_L\downarrow$ is the net radiation loss (R_N), which in this example represents a cooling, and consequently there is a flow of sensible heat from the atmosphere and the lower layers of the soil towards the soil surface.

During the day the situation is more complex because of the incoming short-wave radiation, but the long-wave interactions are similar to those at night, so the net radiation is given by

$$R_N = (R_L\downarrow + R_S\downarrow) - (R_L\uparrow + R_S\uparrow)$$

where the notation is the same as in Figure 2.9.

6 Greenhouse effect

The present surface temperature of the earth represents an energy balance between the visible and near-infrared sunlight that falls on the planet and the middle-infrared thermal emission leaving it. The simplest case is in the absence of an atmosphere, when this equilibrium can be written:

$$0.25(1 - \alpha)S = \varepsilon\sigma T^4$$

where S is the solar constant, α is the albedo of the earth, ε is the mean emissivity of the earth in the middle infrared, T is the effective equilibrium temperature of an earth with no atmosphere, and σ is the Stefan–Boltzmann constant.

Only that part of the earth that faces the sun is illuminated, while the whole of the earth loses infrared radiation to space. The factor 0·25 is the ratio of the area πR^2 that intercepts sunlight to the area $4\pi R^2$ that emits thermal infrared radiation to space. When the best estimates of these parameters are used, a value for T of about 254 K is obtained, and this is far less than the observed mean surface temperature of the earth of 286 to 288 K. The difference is due to the greenhouse effect, in which visible and mean infrared sunlight penetrates through the earth's atmosphere relatively unimpeded, but thermal emission by the ground surface is absorbed by atmospheric constituents that have strong absorption bands in the middle infrared. Thus certain atmospheric gases, principally water vapour and carbon dioxide, absorb a significant part of the outgoing radiation and re-radiate it both upwards and downwards. So the infrared radiation from the earth–atmosphere system to space has three main components. Firstly, there is radiation emitted by the earth's surface at wavelengths to which the earth's atmosphere is transparent —the so called 'window', primarily between 7 and 14 μm. Secondly, some radiation which has been emitted from the surface or from clouds, absorbed by atmospheric gases and re-radiated outward by the same gases; and, lastly, radiation from clouds which themselves may be receiving heat from below. The total amount of infrared radiation to space from the earth–atmosphere system is such that it corresponds to the correct mean equivalent blackbody temperature of about 254 K.

The thermal structure of the earth's atmosphere is therefore influenced by the presence of small quantities of water vapour, carbon dioxide, ozone, trace gases and aerosols. The mean radiation effect of gases, as already explained, is to absorb the upward-moving infrared radiation and to re-radiate it at the local temperature, thus leading to an increase in surface temperature. In contrast, aerosols may either heat or cool the surface, depending on their optical properties for both incident solar radiation and emitted infrared radiation.

Volcanic particles may either reflect solar radiation, which has a cooling effect, or absorb solar energy, which has a warming effect. Pollack *et al.* (1976) concluded that during the first 2 months after a volcanic eruption the two effects either cancel each other out or cause a small net warming of the surface and the stratosphere. Thereafter, the remaining smaller stratospheric dust particles and sulphuric acid aerosols cause a net cooling of the surface. The combined effect over all stages after an eruption is a net cooling of the surface. Now it appears that while stratospheric particles have a net cooling effect for the lower troposphere and the surface of the earth, similar tropospheric particles have a net warming effect. So the net thermal impact of increased atmospheric aerosol loading depends partly on the altitude of the aerosols.

The upward trend in atmospheric carbon dioxide is discussed in detail in Chapter 6, since it will play a major role in future trends of the earth's radiation budget. Variations of the water vapour content of the atmosphere and their influence on the greenhouse effect were discussed in Section 4 of this chapter and are also looked at again in Chapter 5.

It has already been noted in Section 4 of this chapter that the atmosphere contains a large number of trace gases with strong infrared absorption bands. Despite the small amounts of these gases, they can have a significant effect on the atmosphere's thermal structure because they have absorption bands within the 7 to 14 μm atmospheric window which transmits most of the thermal radiation from the earth's surface and lower atmosphere. The concentration of trace gases such as nitrous oxide, methane, ammonia and freons undergo substantial changes because of man's activities. Extensive use of chemical fertilizers and combustion of fossil fuels may perturb the nitrogen cycle, leading to increases in atmospheric N_2O and the same perturbing processes may increase the amounts of atmospheric CH_4 and NH_3. According to Wang *et al.* (1976) doubling the concentrations of N_2O, CH_4 and NH_3 will lead to surface temperature increases of 0·7, 0·3 and 0·1 degree C respectively. So changes in atmospheric trace gases could lead to significant changes in climate via the greenhouse effect.

References

ABBOT, C. G., ALDRICK, W. H. and HOOVER, W. H. 1942: *Annals of the Astrophysical Observatory of the Smithsonian Institution* **6**. Washington.

BAUR, F. 1964: Ist die sogenannte solarkonstante wirklich konstant? (Bericht über die III meteorologische Fortbildungstagung für Grosswetterskunde und langfristige Witterungsvorhersage). *Meteorologische Rundschau* **17**, 19–25.

BRUNT, O. 1941: *Physical and dynamical meteorology*. London: Cambridge University Press.

BUDYKO, M. I. 1974: *Climate and life*. New York: Academic Press.

CADLE, R. D., KIANG, C. S. and LOUIS, J. F. 1976: The global scale dispersion of the eruptive clouds from major volcanic eruptions. *Journal Geographical Research* **81**, 3125–32.

CLIMAP Project Members 1976: The surface of the ice-age earth. *Science* **191**, 1131–7.

DYER, A. J. and HICKS, B. B. 1965: Stratospheric transport of volcanic dust inferred from solar radiation measurements. *Nature* **208**, 131–3.

EDDY, J. A. 1976: The Maunder minimum. *Science* **192**, 1189–202.

ELLIS, A. T. and PUESCHEL, R. F. 1971: Absence of air pollution trends at Mauna Loa. *Science* **172**, 845–6.

ELLISON, M. A. 1956: *The sun and its influence*. London: Routledge and Kegan Paul.

EZER, D. and CAMERON, A. G. W. 1972: Effects of sudden mixing in the solar core on solar neutrinos and ice ages. *Nature, Physical Science* **240**, 180–2.

GATES, W. L. 1976a: Modeling the ice-age climate. *Science* **191**, 1138–44.
  1976b: The numerical simulation of ice-age climate with a global general circulation model. *Journal Atmospheric Sciences* **33**, 1844–73.

KONDRATYEV, K. YA. and NIKOLSKY, G. A. 1970: Solar radiation and solar activity. *Quarterly Journal Royal Meteorological Society* **96**, 509–22.

LAMB, H. H. 1970: Volcanic dust in the atmosphere; with a chronology and assessment of its meteorological significance. *Philosophical Transactions Royal Society* Series A **266**, 425–533.

LOCKWOOD, G. W. 1975: Evidence for solar variability from photometry of planets and satellites. In ZIVIN, H. and WALTER J., editors, *Workshop on the solar constant and the earth's atmosphere*, 181–202. Big Bear Solar Observatory, C.A. Institute of Technology.

MILANKOVITCH, M. 1930: Mathematische Klimalehre und astronomische theorie der Klimaschwankungen. In KÖPPEN, W. and GEIGER, R., editors, *Handbuch der Klimatologie* **1**. Berlin: Teil A.

MONTEITH, J. L. 1973: *Principles of environmental physics*. London: Edward Arnold.

POLLACK, J. B., TOON, D. B., SAGAN, C., SUMMERS, A., BALDWIN, B. and VAN CAMP, W. 1976: Volcanic explosions and climatic change: a theoretical assessment. *Journal Geophysical Research* **81**, 1071–83.

PUESCHEL, R. F., GARCÍA, C. J. and HANSON, R. T. 1974: Solar radiation: effects of atmospheric water vapour and volcanic aerosols. *Journal Applied Meteorology* **13**, 397–401.

ROBINSON, N. 1966: *Solar radiation*. Amsterdam: Elsevier.

RUSSELL, P. B., VIEZEE, W., HAKE, JR. R. D. and COLLIS, R. T. H. 1976: Lidar observations of the stratospheric aerosol: California, October 1972 to March 1974. *Quarterly Journal Royal Meteorological Society* **102**, 675–95.

SCHNEIDER, S. H. and MASS, C. 1975: Volcanic dust, sunspots and long-term climate trends: theories in search of verification. In WMO, *Proceedings of the WMO/IAMAP symposium on long-term climatic fluctuations*, 365–9. Geneva.

SELLERS, W. D. 1965: *Physical climatology*. Chicago: University of Chicago Press.

UK SECTION INTERNATIONAL SOLAR ENERGY SOCIETY 1976: *Solar energy: a UK assessment*. London: UK–ISES.

UNSWORTH, M. H. and MONTEITH, J. L. 1972: Aerosol and solar radiation in Britain. *Quarterly Journal Royal Meteorological Society* **98**, 778–97.

WANG, W. C., YUNG, Y. L., LACIS, A. A. MO, T. and HANSEN, J. E. 1976: Greenhouse effects due to man-made perturbations of trace gases. *Science* **194**, 685–90.

3

The Nature of Surface Climates: Climatic Interaction Models

Energy interactions at a surface depend on the nature of the surface. Thus the albedo of the surface determines how much energy is absorbed while its temperature and moisture state determine how the available energy is utilized. Very high albedo surfaces absorb little short-wave energy and therefore are relatively cool with little available energy. In contrast, temperatures can be high over dry surfaces because the main energy loss is via infrared radiation and sensible heat transfer to the atmosphere. The interactions at surfaces are explored starting with the simplest case first, a dry surface with no atmosphere, and then proceeding to the more complex surfaces covered by moist vegetation.

1 Surface interactions

1.1 Dry surface with no atmosphere

A surface of this nature will assume a very simple energy balance, such that:

$$R_N = R_T(1 - \alpha) - \varepsilon\sigma T^4 - H$$

where

R_N is the net radiation;
R_T is the global radiation;
α is the albedo;
ε is the infrared emissivity;
$\varepsilon\sigma T^4$ is the long-wave radiation loss from a surface at temperature T K;
H is the heat flux into the soil.

Energy fluxes towards the soil surface are taken as being positive. If there is no heat source other than the global radiation, then over long time periods the net radiation R_N will be zero, since the radiative energy gained must equal the radiative energy lost. The flow of heat H into or out of the soil can produce small imbalance in the short term, but in general H is very small compared with the magnitudes of the other energy fluxes. It follows therefore that in general the temperature T will change in close accord with the daily march of incoming radiation, and that it will vary greatly between day and night.

The temperature of the surface for a given global radiation flux depends on its albedo, on its infrared emissivity and on its thermal inertia. The higher the albedo, the more radiation is reflected and the lower the surface temperature. While the emissivity of many natural substances approaches unity it is usually not exactly equal to unity but a few per cent less. It is the surface emissivity which controls the infrared radiation loss by the surface to space. Surfaces with a low emissivity will lose heat by radiation more slowly than surfaces with higher emissivities at the same temperature. Similarly, to lose the same amount of heat by radiation, a surface with a low emissivity will have to be at a higher temperature than a surface with a high emissivity.

The flux of sensible heat into the surface is controlled by the thermal properties of the surface and the subsurface. A natural parameter which expresses the thermal properties of a soil is the thermal inertia $(\rho c \lambda)^{\frac{1}{2}}$, where ρ, c, λ are the density, specific heat and thermal conductivity respectively. If the thermal inertia is large, the subsurface absorbs a large amount of heat during the day and then conducts a large amount of heat to the radiating surface during the night. Temperature variations of the surface will in consequence be moderate. In contrast, soils of poor thermal inertia conduct little heat to the subsurface, and attain high temperatures in the day-time and low temperatures at night.

Obviously there are no surfaces on earth of this type, but the lunar surface is of this particular nature.

1.2 Dry surface with atmosphere present

The surface energy balance equation is the same as before except that the atmosphere is capable of radiating energy R_L and advecting heat S. Thus

$$R_N = R_T(1 - \alpha) - \varepsilon \sigma T^4 + R_L - H - S.$$

The atmosphere will also modify the incoming solar beam by scattering and absorbing it, so not all the radiant energy will come from the direction of the sun. Since the atmosphere also absorbs and re-radiates infrared radiation, there is an infrared flux towards the surface as well as away from it. Two general classes of this particular surface can be found on earth. The first is fairly common and is found in the tropical deserts where it is dominated by the global radiation flux. The second sometimes occurs over dry surfaces at high latitudes in winter and is dominated by the atmospheric heat flux S.

In the tropical deserts there is, because of the clear skies, a large solar radiation input, which leads to high surface temperatures. Since the winds are normally light, horizontal heat transfer by the atmosphere is small, and as a consequence the main heat loss from the surface is by infrared radiation. The infrared radiation losses can be large, leading to relatively small net radiation values despite the large solar radiation input. Thus the surface temperature follows the variations of incoming solar radiation closely and is controlled by it. Great variations of temperature occur under these conditions and numerous stations in the Sahara have recorded maximum temperatures above 45 °C and minimum temperatures below 0 °C. Even higher maximum temperatures would be recorded but for the fact that the albedo of desert surfaces tends to be high and therefore a fair amount of incoming radiation is reflected.

During winter in high latitudes the incoming solar radiation is small, and under these conditions the heat transported by the atmosphere is considerably greater than the incoming solar radiation leading to a dry landscape dominated by the sensible heat flux. Under these conditions, with moderate winds, the surface temperature follows the air temperature which in turn depends on the prevailing synoptic conditions. While wet surfaces of this nature are common, extensive dry surfaces are rare.

1.3 Wet surface with atmosphere present

This is the most common type of surface found in nature and the surface energy balance may be described by the equation in the previous section together with the addition of the energy lost during evaporation (LE_a):

$$R_N = R_T(1 - \alpha) - \varepsilon\sigma T^4 + R_L - H - LE_a - S$$

where L is the latent heat of vaporization, and E_a is the actual evaporation. A wet surface in this context is one from which the evaporation is controlled solely by the prevailing radiational and meteorological conditions, that is to say the evaporation is independent of the rate of water supply. Such a surface could be one which is physically wet such as an ocean or a landscape after rain, or it could be one in which evaporation is occurring freely from water moving through plants (transpiration).

Evaporation from bare soil soon becomes limited because the surface layers become dry and under strong radiative conditions the rate of evaporation exceeds the rate at which water can move upwards to the soil surface. Vegetation, by sending roots deep into the soil, is able to draw water from some depth, and this may allow it to evaporate freely when the surrounding bare soil surfaces are dry. As the soil dries the remaining water is held at increasing tensions until a point is reached when it is so tightly bound in the soil that it is unable to enter the plant roots and evaporation will become insignificant. Though at this point the soil may still hold a considerable amount of water, it is not available for evaporation and marks the lowest moisture content at which evaporation is significant. It is known as the permanent wilting point of the soil. When the soil is saturated it is said to be at field capacity, that is to say the soil profile retains the maximum amount of water against gravity, any excess water draining away. Neither the wilting point nor the field capacity represent intrinsic properties of the soil and can only be arbitrarily defined, but nevertheless they have great potential value and can be used to obtain an idea of the storage capacities of soils. The difference between the soil water content at field capacity and at wilting point is known as the available water content. Clearly, the available water content will be both a function of the rooting depths of the plants and of the soil type. Some typical values are given in Table 3.1. Thus it is seen that the water holding capacity of the landscape is both finite and relatively small, and that it can soon be exhausted by evaporation.

While the soil forms the main moisture store in the landscape, there are other important stores of somewhat smaller magnitude. These are shown in Figure 3.1 and consist of water held on plant surfaces, water held within plants, and water held on the soil surface in the form of small puddles.

After rainfall, evaporation takes place from water held on plant surfaces and on the soil surface. Under these conditions evaporation will be unrestricted by the soil moisture state and will be at the maximum rate possible for the given meteorological conditions. The combined process of evaporation

Table 3.1 Estimated maximum available water content of various crops (*data from Grindley, 1972*).

Crop	Maximum available water content (mm)
Wheat	200
Oats	200
Turnips, swedes and fodder beet	150
Orchards, grown commercially	225
Bare fallow	25
Temporary grasses	100
Permanent grass	125
Rough grazing	50
Permanent woodland	250

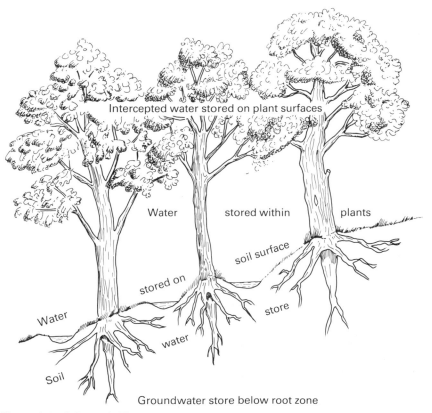

Intercepted water stored on plant surfaces

Water stored within plants

soil surface

stored on

Water store

water

Soil

Groundwater store below root zone

Figure 3.1 Schematic illustration of the main water stores in the landscape.

from water surfaces and transpiration from vegetation is often known as evapotranspiration. Potential evapotranspiration is the maximum amount of water vapour that can be added to the atmosphere under the given meteorological conditions from a surface covered by green vegetation with no lack of available water.

As the soil dries, so the remaining soil moisture is held in the soil at increasing tensions. Eventually the tension at which water is held in the soil will become so great that evapotranspiration will fall below the potential value. Now it has been found by various investigations that as the soil dries evapotranspiration is first of all near the potential rate, but that ultimately it falls below the potential rate. Priestley and Taylor (1972) claim that observations over a considerable variety of crops are consistent with the view that when evapotranspiration ceases to be at the full potential rate, the ratio of actual to potential evapotranspiration rate falls off linearly with increasing soil moisture deficit and approaches zero when the soil moisture deficit is about 50 mm more than it was when the surface first ceased to behave as a saturated one. The process is illustrated in Figure 3.2. Soil moisture deficit is defined as the amount of rainfall or irrigation required to restore the soil to its field capacity.

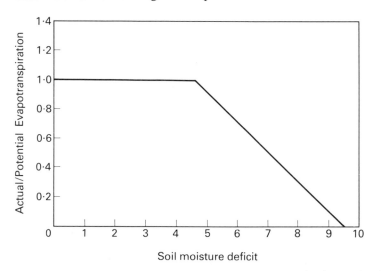

Figure 3.2 The ratio of actual to potential evapotranspiration under drying conditions with a resulting increase in the soil moisture deficit. The potential evapotranspiration is assumed to be at a constant rate throughout. (Units: cm of rainfall.)

1.3.1 *Evaporation*
In meteorology, evaporation is the change of liquid water or ice to water vapour, and it proceeds continuously from the earth's free water surfaces, soil, snow and ice-fields. Transpiration is the process by which the liquid water contained in the soil is extracted by plant roots, passed upward through the plant, and discharged as water vapour to the atmosphere; the rate of transpiration is highest during the day and falls almost to zero during night hours.

Evapotranspiration is the combined process of evaporation from the earth's surface and transpiration from vegetation. As already noted, potential evapotranspiration is the maximum amount of water vapour that can be added to the atmosphere under the given meteorological conditions from a surface covered by green vegetation with no lack of available water.

The evapotranspiration from a land surface depends primarily on the radiant energy supply to the surface, but it may be limited by the saturation deficit of the air and the rate of movement of water to the evaporating surface. The saturation deficit is the difference between the actual vapour pressure of a moist air sample at a given temperature and the saturation vapour pressure corresponding to that temperature. For potential evapotranspiration (E_t) the two important terms are the surface energy balance and the saturation deficit and these were combined in one equation by Penman (1948):

$$E_t = \frac{\Delta(R_N - G)/L + \gamma E_a}{\Delta + \gamma}$$

where all the terms are as before except for the following:

G   is the heat flux into the ground;
γ is the psychrometric constant (about 0·66);
Δ is the slope of the temperature–vapour pressure curve at the air temperature;

and E_a is the aerodynamic evaporation, which depends on both the saturation deficit and the wind speed and may be estimated from equations of the form:

$$E_a = f(\bar{u})(e_s - e_a)$$

where e_s and e_a are the saturated and actual vapour pressures respectively; and $f(\bar{u})$ is an empirically derived function of the wind velocity, usually given in the form:

$$f(\bar{u}) = a(b + c\bar{u})$$

where $\bar{u}$ is the time-averaged wind velocity measured at a standard height and a, b and c are constants.

Values Δ are such that the aerodynamic term E_a determines the evapotranspiration E_t at low temperatures, like those found in Europe in winter, while the term R_N determines the evapotranspiration at high temperatures, like those found on sunny days in Europe in summer or in the tropics.

1.3.2 *Water storage on vegetation*

Much of the evaporation water loss from the landscape is from water held on plant surfaces. It is therefore necessary briefly to consider this particular water storage.

Vegetation may intercept water falling in the form of snow, hail or rain, as well as droplets of water in low cloud and fog which would not otherwise be precipitated. In the initial stage of interception by a dry canopy, much of the water is retained. There appears, however, to be a fairly well defined storage capacity for any given vegetation canopy, and when this is exceeded, further intercepted water either drips from the canopy or runs down the stems. The combination of the water which drips from the canopy and falls directly through gaps is usually called throughfall and the sum of throughfall and stemflow is net precipitation. The difference between gross precipitation and net precipitation is often called interception loss, the loss representing water which is evaporated from plant surfaces.

The storage capacity of a canopy can be estimated approximately as the constant *C* in the regression equations for individual storms:

net precipitation = *b*(gross precipitation) − *C*

The coefficient *b* will be unity only if the storms are large enough to wet the canopy completely, that is consisting of continuous falls with negligible evaporation, separated by periods long enough for all the surface stored water to be evaporated, some typical storage values are given in Table 3.2.

Table 3.2   Interception storage capacities (*after Rutter, 1975*).

Vegetation		C (mm)
Coniferous forest		
Pinus sylvestris		1·6
Picea abies		1·5
Pseudotsuga menziesii		2·1
Pinus nigra		1·0
Deciduous forest		
Carpinus betulus	summer	1·0
	winter	0·6
Old *Quercus robur* coppice	summer	1·0
	winter	0·4
Ericaceous		
Calluna vulgaris		2·0
Herbaceous		
Zea mais		0·4–0·7
Mixed grasses and legumes		1·0–1·2
Lolium perenne, 10 cm high		0·5
48 cm high		2·8
Molinia caerulea		0·7
Pteridium aquilinum		0·9

Following Penman's equation for evaporation, transpiration from vegetation may be expressed by:

$$LE_t = \frac{\Delta(R_N - G) + \rho C_p \delta e / \Gamma_a}{\Delta + \gamma(1 + \Gamma_c / \Gamma_a)}$$

The previously undefined parameters are as follows:

ρ is air density;

C_p is the specific heat of air at constant pressure;

δe is the saturation deficit, i.e. $(e_s - e_a)$;

Γ_a is the boundary layer resistance of the canopy (depends partly on roughness of canopy);

Γ_c is the surface (biological) resistance of the canopy (a measure of the bulk stomatal resistance of a particular plant community).

Boundary layer and surface resistances may be defined using Ohm's law in electricity as a direct analogue. Ohm's law states that the electrical resistance of a wire is equal to the potential difference between its ends divided by the current flowing through it: i.e.

$$\text{electrical resistance} = \frac{\text{potential difference}}{\text{current}}$$

The analogue relationship is obtained by replacing 'potential' by concentration (meaning amount per unit volume) and 'current' by flux (amount per unit area per unit time), so that:

$$\text{boundary layer or surface resistance} = \frac{\text{concentration difference}}{\text{flux}}$$

The evaporation rate of intercepted water is given by:

$$LE_i = \frac{\Delta(R_n - G) + \rho C_p \delta e / \Gamma_a}{\Delta + \gamma}$$

Assuming that the albedos of wet and dry canopies are not materially different, then

$$\frac{E_t}{E_i} = \frac{\Delta + \gamma}{\Delta + \gamma(1 + \Gamma_c/\Gamma_a)}$$

When Γ_c and Γ_a are of similar size, as they may be in herbaceous communities, then the rate of evaporation of intercepted water will not much exceed the potential transpiration rate in the same conditions. Indeed, for grass it is often assumed that they are the same. In contrast, when Γ_c is an order of magnitude greater than Γ_a, as it appears to be in coniferous forest, then intercepted water will be evaporated at 2 to 5 times the current transpiration rate.

During winter net radiation values are small or negative but the aerodynamic evaporation is still significant, and can lead to potential evapotranspiration rates of 2 or 3 mm day^{-1}. This value is equal to or larger than the interception capacities of many forests, so large amounts of water can be evaporated from forests under conditions of light rainfall. Under these conditions the evapotranspiration from a rough forested landscape can exceed that from an equivalent grassland landscape.

1.3.3 *The Bowen ratio*

The study of the energy balance of natural surfaces falls into two stages. First, there is the study of the radiation balance, which leads to an estimation of the available net radiation. Secondly, the net radiation has to be divided amongst the sensible heat flows to the soil and the atmosphere, and the latent heat flow to the atmosphere, to produce the full energy balance. The ratio of the sensible heat flow to the atmosphere to the latent heat flow is known as the Bowen ratio, β, and can be written as:

$$\beta = \frac{\text{sensible heat loss to atmosphere }(C)}{\text{latent heat loss to atmosphere }(LE_t)}$$

In the absence of atmosphere advection, β can vary between $+\infty$ for a dry surface with no evaporation to zero for an evaporating wet surface with no sensible heat loss. If there is atmospheric heat advection, β may become negative indicating a flow of heat from the atmosphere to the surface.

Both the sensible and latent heat fluxes in the vertical can be written in the convenient, almost symmetrical, forms:

$$C = -\rho C_p K_H \frac{\partial T}{\partial z}$$

$$LE = -\rho \frac{C_p}{\gamma} K_V \frac{\partial e}{\partial z}$$

where

ρ is the density of moist air;
C_p is the specific heat of air at constant pressure;
K_V and K_H are the eddy diffusivities for water vapour and heat respectively;
$\frac{\partial e}{\partial z}$ and $\frac{\partial T}{\partial z}$ are the vertical gradients of vapour pressure and temperature respectively;
L is the latent heat of vaporization of liquid water.

The thermodynamic value of the psychometric constant, γ, is given by:

$$\gamma = \frac{C_p p}{0 \cdot 621 L}$$

γ becomes $0 \cdot 66$ if the following typical values are introduced into the above equation:

$C_p = 0 \cdot 240 \text{ cal } {}^\circ C^{-1} \text{ g}^{-1}$
$L = 585 \text{ cal g}^{-1}$
$p = 1000 \text{ mbar } (p = \text{pressure}).$

Therefore, from the above equations it is possible to write for the Bowen ratio:

$$\beta = \gamma \frac{\partial T}{\partial e} \frac{K_H}{K_V}$$

Often the convenient simplification is made that $K_H = K_V$, giving:

$$\beta = \gamma \frac{\partial T}{\partial e}$$

Priestley and Taylor (1972) have developed further the idea of the Bowen ratio. To a good approximation the specific humidity q is given by:

$$q = 0.622\frac{e}{p}$$

It is therefore possible to replace e by q in the relationship for the Bowen ratio, and from the equation for γ obtain:

$$\beta = \frac{C_p}{L}\frac{\partial T}{\partial q} = \frac{C_p}{L}s$$

This is a useful relationship because the profiles of specific humidity and temperature are often similar if the surface is saturated. Priestley and Taylor also introduce a term α, where

$$\frac{LE_t}{LE_t + C} = \alpha\frac{s}{s + \gamma}$$

For free evaporation from a uniform saturated surface, with no advection effects, they found that $\alpha = 1.26$. If α is taken as 1.26, it follows that:

$$\beta = \frac{C}{LE} = \left(1 - 1.26\frac{s}{s + \gamma}\right)\Big/\left(1.26\frac{s}{s + \gamma}\right)$$

which has been found to be a function of the surface temperature. Values of β calculated on this basis are given in Figure 3.3. This figure suggests that over a well watered surface the Bowen ratio decreases as the temperature increases, or the proportion of available energy going into latent heat increases. At

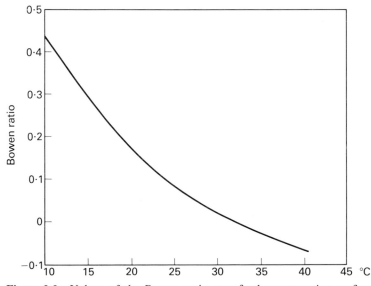

Figure 3.3 Values of the Bowen ratio at a freely evaporating surface for various surface temperatures (T_0). This diagram only applies to a freely evaporating surface, for if the surface becomes dry, the value of the Bowen ratio will increase to above that appropriate for the temperature T_0.

temperatures above 32 °C the sensible heat flow becomes negative implying a flow of heat from the air to the evaporating surface. This would seem to suggest that the highest temperature which can be reached over a freely-evaporating surface with the net radiation values experienced on the earth's surface is about 32 °C. There is evidence to support this particular view. Priestley (1966) examined the average daily maximum temperature for each month reported by island observing stations and by land stations after periods of heavy rain. His conclusion was that, in the radiation climates that actually exist in nature, air temperatures over a well watered surface do not rise above 32–34 °C. Similarly, Linacre (1967) examined the relationship between the temperature of freely evaporating leaves in bright sunshine and of the air and reached the conclusion that leaves are hotter than the air up to about 33 °C and, above that, they are cooler.

The previous discussion only applies if the evaporation is not restricted and there is no advection of heat. Advection effects become particularly important if a warm air current flows across a cold surface. Under these circumstances heat is transferred from the atmosphere to the surface and then used for evaporation. Thus the wet surface allows some of the sensible heat of the atmosphere to be converted into latent heat. Figure 3.4 illustrates this particular effect for an air current flowing over the sea. It is seen that when the air–sea temperature difference becomes negative, the Bowen ratio also becomes

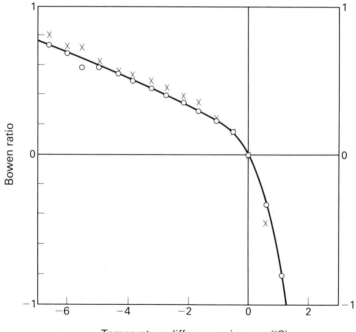

Figure 3.4 Bowen ratio as a function of the air–sea temperature difference. Circles, wind force 4 Beaufort; crosses, wind force 8 Beaufort (*from Roll, 1965*).

negative. Thus evaporation rates into a cold air current flowing over a warm sea may be unexpectedly high. Wind velocity appears to have little influence on the Bowen ratio.

The Bowen ratio will also rise above that appropriate for a freely-evaporating surface if the evaporation is limited for some reason. This may be because the soil becomes dry and the vegetation is under stress, or because the plants themselves limit evaporation.

Some typical climatological values of the Bowen ratio are shown in Table 3.3. The values are for grassland on the eastern slopes of the Pennines, and it

Table 3.3 Average monthly Bowen ratios for a typical year over grass in the eastern Pennines.

Month	Theoretical Bowen ratio for average surface temperature	Actual Bowen ratio
January	0·80	− 4·25
February	0·79	− 4·08
March	0·66	0·09
April	0·69	0·46
May	0·48	0·41
June	0·32	0·53
July	0·31	0·67
August	0·31	0·33
September	0·38	0·27
October	0·58	− 0·51
November	0·74	19·2
December	0·76	− 9·54

is assumed that evaporation rate from the grass is not restricted unless the soil moisture becomes limiting. The values of Bowen ratio, evaporation, etc., were calculated using a model of the type described in Chapter 6. Average monthly Bowen ratios were calculated from both the average temperature and from the estimated actual evaporation and sensible heat fluxes.

Theoretical Bowen ratios calculated from temperature considerations show only a gradual change from winter to summer, being highest in January and lowest in July and August. The variation of the actual values is far more complex. Average actual Bowen ratios for the winter months of December, January and February are negative, as is the value for October. This suggests that there is, in winter, a flow of sensible heat from the atmosphere to the surface, and that evaporation rates are controlled by atmosphere properties such as saturation deficit and wind speed. The very high value of the Bowen ratio for November is due to a period at the end of the month when temperatures fell rapidly to below freezing and is highly suspect. During the spring there is a flow of heat into the soil and this tends to depress the Bowen ratio values.

If the November actual Bowen ratio is ignored, the maximum actual values are found in June and July, when the theoretical value is approaching its minimum. This is because a lack of soil moisture during the summer produces some stress in the vegetation and limits evaporation to below the potential value. At the end of June the soil was dry, and the average actual Bowen ratio for the first five days of July rose to 3·4 as compared to a value of 0·17 calculated assuming evaporation at the potential rate. The corresponding theoretical value of the Bowen ratio was 0·23. During the next five days it rained heavily and the actual average Bowen ratio fell to 0·53. Both the Bowen ratio calculated assuming potential evaporation and the theoretical value were equal at 0·3. It is therefore clear that the actual value of the Bowen ratio is partly determined by the moisture status of the soil.

Bowen ratio values can also rise to high values if the plant itself restricts transpiration from its leaves and examples of this are found in coniferous forests. According to Jarvis *et al.* (1976), measurements over coniferous forests fall into two groups. For most coniferous forest sites, irrespective of species, the day-time Bowen ratio of a dry canopy varies between 0·1 and 1·5. When the canopy is wet, and evaporation rather than transpiration is occurring, it is reduced to between $-0·7$ and $+0·4$. However, at two United Kingdom sites, Thetford (52°25′ N, 00°39′ E) and Fetteresco (56°38′ N, 02°24′ W), much larger values of β have been found consistently, and similar large values are found occasionally in other extensive records. At Thetford, the Bowen ratio rises to a fairly steady value of between 1 and 4 soon after sunrise; in the afternoon, it may decrease or, in conditions leading to stomatal closure, may climb to even higher values. At Fetteresco, it may reach 2 to 3 in the middle of the day, but declines again in the afternoon.

Monteith (1965) generalized the Penman equation for evaporation from a short crop in the form:

$$LE_t = \frac{\Delta R_N + C_p \rho \delta e / \Gamma_a}{\Delta + \gamma(1 + \Gamma_c/\Gamma_a)}$$

where δe is the atmosphere water vapour saturation deficit (mbar), Γ_a is the boundary-layer resistance of the canopy to mass transfer, and Γ_c is the surface resistance to water vapour transfer.

Combining the above equation with

$$R_N = LE_t + C$$

and

$$\beta = \frac{C}{LE_t}$$

produces

$$\beta = \frac{1 + \gamma(1 + \Gamma_c/\Gamma_a)/\Delta}{1 + C_p \rho \delta e / \Delta R_N \Gamma_a} - 1$$

Following Monteith (1965) and Stewart and Thom (1973), a 'climatological resistance' is defined as:

$$\Gamma_i = \frac{\rho C_p \delta e}{\gamma R_N}$$

Therefore,

$$\beta = \frac{\Delta/\gamma + 1 + \Gamma_c/\Gamma_a}{\Delta/\gamma + \Gamma_i/\Gamma_a} - 1$$

It is apparent that β is approximately proportional to the surface resistance, Γ_c, when Γ_c is large in relation to the boundary-layer resistance, Γ_a; and similarly that β is approximately proportional to R_N and inversely proportional to δe when Γ_i is large in relation to the boundary-layer resistance, Γ_a. The very large values of the Bowen ratio at Thetford almost certainly result from stomatal closure increasing Γ_c, whereas the generally high values in the middle of the day at Fetteresco are largely the result of the prevailing small vapour pressure deficits in an oceanic climate. Jarvis *et al.* comment that Bowen ratios listed for other coniferous forests are generally smaller than the British examples because the sites are nearly all much more continental, and also because measurements on clear warm summer days with large water vapour deficits are frequently reported.

1.4 Radiative and advective landscapes

It will have become clear from the discussions in the previous sections that the energy transfers taking place at landscape surfaces may be dominated by radiative or advective effects. On clear, sunny days the radiative fluxes are considerably greater than the advective fluxes, and there is therefore a large transfer of energy to the atmosphere. In contrast, when the radiative fluxes are small, the energy exchanges at the surface are dominated by the advective heat flux and the vapour pressure deficit. High temperatures, such as those found in the tropical zone, are usually the result of solar heating, advection effects tending to be more important at low temperatures. The slope of the saturated vapour pressure/temperature curve (Figure 3.5) increases with increasing temperature, so even with relatively low vapour pressures the saturation deficit of the air is normally low at low temperatures, leading to restricted evaporation. At higher temperatures, the saturation deficit of the air tends to be greater, and evaporation is controlled by the available net radiation.

In the case of a landscape dominated by advection the prevailing temperatures will be those of the prevailing local air mass, radiative effects being of little importance. This particular type of landscape is found in most middle- and high-latitude locations under moderate to strong winds in winter. If the wind is calm or light, radiative effects may be important, and under clear skies very low temperatures may be observed at high latitude sites in winter. Winter surface radiation balances are largely negative over both Eurasia and North America in winter, so unless temperatures are to fall to very low values the energy loss to space must be replaced by atmospheric heat advection. Even in summer advective effects can be important at middle and high latitudes under cloudy and windy conditions. Rutter (1975) considers that, in England, the temperatures and saturation deficits differ much less between winter and summer on rainy days than they do on dry days and that wind speeds tend to be higher in winter. It follows from this that the evaporation of intercepted water on vegetation differs little between British summers and winters.

Mean temperature charts for Europe show that the isotherms have a zonal trend in summer but are more meridional in winter, and this partly reflects

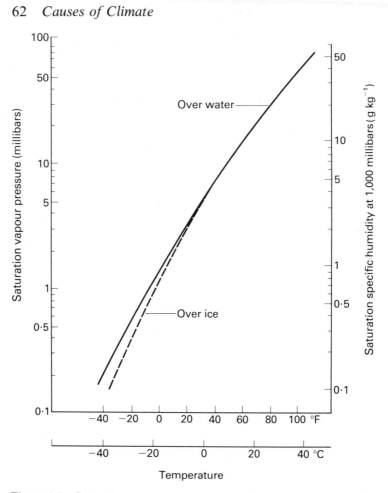

Figure 3.5   Saturation vapour pressure at various temperatures. Below 0 °C there are two curves, one for supercooled water, the other for ice.

the difference between advection-dominated and radiation-dominated land-scapes. In winter, the net radiation balance of both the continental surface and the earth–atmosphere system over the continent is negative, and the radiation deficit is made up by the advection of heat from the west and south. Therefore, the progressive cooling of the air masses as they pass over the cold continent from the west is probably the explanation for the trend of the January isotherms. In summer, the net radiation balances over the continent are positive and the advection of warm air is not of great impor-tance, except perhaps in the south. The summer net radiation values fall towards the north and this is reflected by the zonally decreasing tempera-tures. Winter temperatures in Europe therefore tend to reflect air mass sources and characteristics, and in the long term the temperature of the North Atlantic Ocean. A less vigorous westerly circulation with greater easterly and northerly components would cause generally lower winter temperatures. European summer temperatures tend to reflect much more the degree of atmospheric cloudiness, windy wet summers being cooler than sunny ones.

Mountains provide excellent examples of the influence of atmospheric advection on local climatology. Western European uplands are interesting because they are abnormally cool as compared with those further east, and in particular, landscapes resembling those of the tundra may be found at elevations as low as 600–800 m in Britain. This is partly a result of the dominance of maritime polar air in north-west Europe, with its associated strong winds, frequent low cloud, and steep lapse rate. Under these conditions temperatures observed over British hills are low and close to those in the free atmosphere. Further east, with clearer skies and lighter winds, the positive radiation balance in summer gives rise to relatively high temperatures in the uplands and causes the cultivation limit to increase in height inland. The summer temperatures experienced on temperate-latitude mountains depend largely on how effective the mountains are in creating their own local climates. If the winds are light and the uplands are very large and massive, they will be able, under the influence of the positive net radiation in summer, to create a local climate which is largely independent of that in the free atmosphere. Under these conditions, differences in radiation receipts resulting from topographic factors are reflected in soil and air temperatures, snow cover duration, soil moisture and consequently vegetation.

2 Examples of natural surfaces

A large range of surfaces are found in nature and it is useful to discuss some of their more interesting properties in the light of previous discussion.

2.1 Oceans

Over half of the solar radiation reaching the earth's surface is absorbed by the oceans. This solar radiation, along with surface wind stresses and variations in salinity, is the ultimate energy source for a variety of physical processes in the ocean which are of some climatic importance. The energy balance of the oceans corresponds to that of the wet surface discussed earlier, and is shown in more detail in Figure 3.6.

Some of the radiation that reaches the sea surface is reflected back into the atmosphere. This reflectance, defined as the ratio of the reflected to the incident radiance, is not identical with the albedo of the sea surface. This is because the albedo is the ratio of all the short-wave radiation that leaves the sea to the incident irradiance, and the former includes not only light reflected at the surface, but also light which has been scattered upwards from within the water mass. The reflectance from the sea surface increases with the zenith angle, thus when the sun is low in the sky, more light is reflected and less penetrates the water, than when the sun is high. Similar comments apply to that part of the diffuse sky radiation which comes from

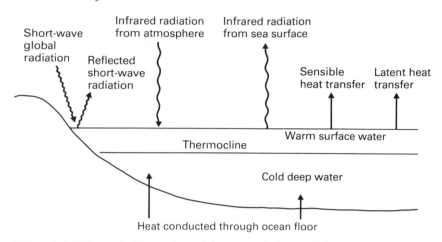

Figure 3.6   Schematic illustration of the energy balance of the ocean.

regions close to the horizon. So clouds increase the reflectivity when the sun is high because their presence then increases the fraction of radiation that comes from near the horizon. In contrast, when the sun is low, clouds allow more light to penetrate the water because they then increase the light scattered vertically downwards. The reflectance of the sea is also influenced by other factors. The white-caps on waves have a diffuse reflectivity which is generally much larger than any flat water surface. Similarly the albedo of sea ice can vary from less than 0·4 for dirty melting old ice to as much as 0·9 for ice covered with new snow. Even in the cleanest ocean water, only 1 per cent of the light penetrates below 100 m, so more than 97 per cent of the ocean volume is in perpetual darkness.

The infrared upward radiation from the sea surface is slightly less than the blackbody radiation. This is due partly to reflection at the interface and partly to its originating at some depth where the water temperature is different from that at the surface itself. Values of the infrared emissivity of the sea have been variously estimated between 0·90 and 1·00, depending on sea state and surface contamination.

Even in the absence of clouds, the downward infrared radiation depends on the distribution of temperature and humidity throughout the vertical. However, a relatively large fraction of the downward stream originates in the lowest 20 m. Now high sea surface temperatures are associated almost invariably with relatively high humidity values in the lower atmosphere. As a result, climatological averages of the net infrared radiation flux as observed in nature or as calculated theoretically vary strikingly little with the temperature. This is demonstrated by Table 3.4, where it is seen that a fall in the relative humidity of the air will lead to a substantial increase in the net infrared radiation loss. Thus the net infrared radiation at the sea surface under clear skies is controlled by both the sea surface temperature and the atmospheric humidity. The net infrared radiation loss will be further modified by the presence of clouds.

Table 3.4 The estimated net infrared heat loss from the surface with cloudless condition (cal cm^{-2} min^{-1}) (*after Budyko, 1956*).

Temperature (°C)	Humidity of the air (mm of Hg)										
	1	2	3	4	5	6	7	8	10	12	15
−5	0·14	0·13	0·12								
0	0·15	0·14	0·13	0·12							
5	0·16	0·15	0·14	0·13	0·13	0·12					
10	0·17	0·16	0·15	0·14	0·14	0·13	0·12	0·11			
15		0·17	0·16	0·15	0·15	0·14	0·13	0·12	0·11	0·10	
20			0·17	0·16	0·16	0·15	0·14	0·13	0·12	0·11	
25				0·17	0·17	0·16	0·15	0·14	0·13	0·12	0·10
30					0·18	0·17	0·16	0·15	0·14	0·13	0·11

The penetration of heat into the temperate oceans represents a much more complicated process than on land where only the process of physical heat conduction acts. During heating in spring, the temperature increase is restricted to only a thin layer which becomes less dense. Therefore, the stratification becomes more stable, and the development of a thermocline is initiated. A thermocline is an ocean layer in which the rate of decrease of temperature with increasing depth is a maximum. It is a stable region in which the vertical mixing of water is strongly inhibited. In summer, a 'seasonal thermocline' is formed in most middle-latitude oceans, its depth varying from about 15 m to about 50 m. During the autumn when cooling begins, the surface water becomes denser, and vertical thermal convection becomes effective. The thermocline diminishes until the temperature differences disappear and winter conditions are established. If salinity differences do not contribute to the density stratification, the water column from the surface to the bottom becomes isothermal in shallow seas such as the North Sea. In temperate latitudes, the isothermal layer extends to approximately 200–300 m in the open ocean. In some ocean areas at high latitudes the isothermal layer can be traced to great depths and leads to the formation of cold ocean bottom water. Normally a permanent thermocline exists below the seasonal thermocline at a depth of some hundreds of metres in low- and middle-latitude oceans. The seasonal thermocline is absent or very weak in the tropical oceans.

The total mass of the oceans is about 280 times that of the atmosphere, while their heat capacity is nearly 1,200 times larger. Although the annual radiation cycle affects only a small part of the water mass, the thermal inertia is strong enough to prevent large or fast temperature variations. This has a dominating influence on the whole terrestrial climate, and is particularly strong in maritime regions. By virtue of both their mechanical and thermal inertia, the oceans tend to play the role of a fly-wheel in the air–sea system.

Marked variations from average in sea surface temperatures can be caused by a variety of factors, and one example is discussed for the equatorial Pacific in later chapters. The flux of sensible and latent heats from the oceans is largest over the tropical half of the globe. This energy is advected horizontally by the atmosphere, until ultimately much of it is lost into space by infrared radiation in higher latitudes. Evaporation from the tropical oceans feeds the monsoon rains and most of the great rivers that drain the continents. All these events may be affected significantly by variations in sea temperature, in atmospheric stability and in the resulting air–sea interactions.

Over monthly periods, the fluctuations of the sea temperature tend to contribute very little to the variance of the sea–air temperature differences; most of it can be attributed to fluctuations in the atmosphere. Sea-temperature variations become more important over longer periods such as whole seasons. In temperate latitudes the effect of sea-temperature variations on the sea–air temperature difference becomes relatively more important in summer. This is due to the absence of large horizontal temperature contrasts in the atmosphere during that season. At the same time, the sea-surface temperature variations are relatively large in summer because that is the season when the surface mixed layer has the smallest depth.

In the long term, large climatic effects are associated with changes in the location or temperature of ocean currents. Thus temperature anomalies in the Pacific Ocean are claimed by Namias (1969, 1970) to be connected causally with the weather pattern over North America. Anomalously warm sea surface areas are also areas of preferred cyclone formation, which in turn change the phase of the quasi-stationary long planetary waves in the upper atmosphere over the Pacific–North American sector.

2.2 Grasslands

Grassland is both a relatively common and relatively simple vegetation type. The water balance of grass in particular is simple in that moist grass transpires at a rate which is near to the maximum value for a moist surface under the given meteorological conditions. This is in contrast to coniferous forest where transpiration rates are severely limited.

Energy interactions over grassland surfaces tend to be relatively simple. It has been found from experiments by Monteith and Szeicz (1961) that there is a linear relationship between the net radiation (R_N) received by a grassland surface and the global radiation (R_s) during cloudless summer days. Monteith and Szeicz suggested that the relationship was described by:

$$R_N = \frac{1 - \alpha}{1 + \beta} R_s + L_0$$

where α is the short-wave albedo; β is a heating coefficient defined as the increase in long-wave radiation loss per unit increase in R_N; and L_0 is the net radiation as $R_s \rightarrow 0$.

Values of β depend largely on the nature of the surface and slightly on atmospheric properties. If the surface is dry and evaporation is greatly restricted, much of the available net radiation will form sensible heat and the surface heating will be large. In contrast, if it is wet the surface heating will be small because much of the available energy will be used for evaporation. Thus β is a function of surface moisture and also the Bowen ratio. Since the Bowen ratio is partly dependent on surface temperature, β is probably also slightly temperature dependent. Surface heating also depends on the stability of the atmosphere and could be increased by the presence of an inversion. Monteith and Szeicz suggest that β is close to 0·1 for agricultural crops and natural vegetation which completely covers the ground and are never short of water. If transpiration is physiologically restricted or if the ground cover is incomplete, β may lie between 0·1 and 0·2, and for very dry soil between 0·3 and 0·4. For clear sky conditions, L_0 is of the order of 70 W m^{-2}.

The albedo of grassland is not simple since the upward reflected radiation is composed of both reflected and multiple scattered radiation by plant organs and by the ground surface. The diurnal variation of grassland albedo on a sunny day shows the typical bowl-shaped curve illustrated in Figure 3.7. Because of the orientation of grass leaves, albedos are always least around noon on sunny days and increase with decreasing solar elevation.

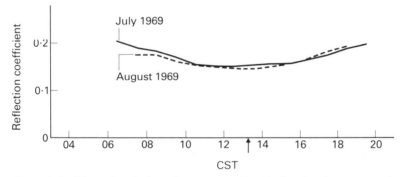

Figure 3.7 Diurnal variation of mean monthly albedo of prairie at Matador for the months of July and August 1969. The arrow shows the approximate time of solar noon (*after Ripley and Redmann, 1976*).

The albedo of bare soils depends mainly on their organic matter content, on water content, particle size and angle of incidence. The reflectivity of a soil sample decreases as it gets wetter, mainly because radiation is trapped by internal reflection at air–water interfaces formed by the menisci in soil pores. The albedo of a stable soil could therefore be used to monitor the water content of the surface layer.

Figure 3.8 shows the variation of global and net radiation over short grass at Kew during 1976. The net radiation is measured over short grass, but the water table tends to be high at Kew so there is little sign of the long summer drought of 1976. Under very dry conditions the net radiation decreases as the infrared radiation increases with rising surface temperatures. There are some signs of this in August when increasing global radiation is accompanied by steady or falling net radiation values. The drought ended at the beginning of September, and the cloud and rain is reflected in the

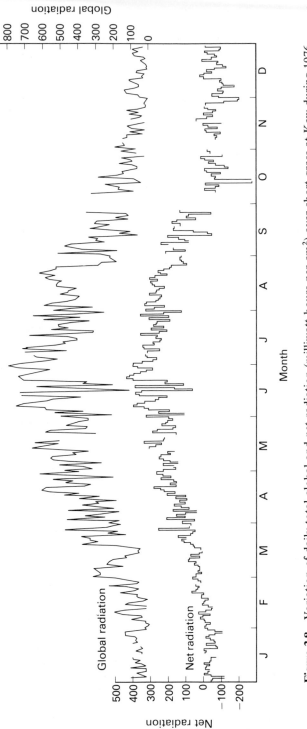

Figure 3.8 Variation of daily total global and net radiation (milliwatt-hours per cm²) over short grass at Kew during 1976 (*data provided by Meteorological Office, Bracknell*).

decreased global radiation values. During the winter months the net radiation is mostly negative, which is typical of most temperate-latitude sites.

2.3 Deserts

The energy balance of a desert differs from that of the surrounding vegetated areas in that the surface is dry and therefore there is little or no evaporation. Figure 3.9 shows the radiation budget of the surface–atmosphere system from Nimbus III data during the period 16–30 June 1969. The balance is positive almost everywhere to the north of 10° S with a maximum in the cloudless zones of the northern areas of the subtropical oceans. Negative radiation balances are observed over the desert regions of north Africa and the Middle East. There are several reasons for these negative balances during summer over the northern deserts. Because the surface is dry there is no evaporation and surface temperatures become high with an associated large infrared radiation loss. The lack of clouds, and of water vapour in the air, allows unusually large amounts of radiation to escape from the surface to space. Lastly, the albedo of desert tends to be high, probably between 30 and 40 per cent, as compared with vegetated surfaces. Because of this negative radiation imbalance, energy must be imported into the desert region to fill the deficit and maintain the temperature of the surface. This extra energy is supplied by the atmospheric circulation. The air that descends into the deserts originally rose near the equator, and supplies the extra heat from the equatorial regions. It is therefore necessary to discuss energy transport in the atmosphere.

The energy content of one gram of moist air may be written as follows:

$$\text{total energy content } (Q) = \text{latent heat content } (Lq) + \text{sensible heat content } (C_pT) + \text{potential energy content } (gz)$$

In the following discussion it is assumed that one gram of moist air is under discussion, so mass will often be omitted. The first component of the total energy content is the latent heat term, which consists of the latent heat of vaporization (L) multiplied by the mass of water vapour (q). Sensible heat is the product of the specific heat (C_p) and the temperature (T) in degrees Kelvin, and is contained in the second term. The atmosphere is continually losing heat, in the form of radiation to space, at a rate of about $\frac{1}{4}$ per cent per day of the total energy content; this represents a cooling of about 1 or 2 degree C per day. Sensible heat is gained from the surface and by the release of latent heat from condensing water vapour. Potential energy exists in the unit parcel by virtue of its position above the earth's surface. It is the product of the height z above sea-level and the force of gravity g. If the parcel sinks slowly, the potential energy must decrease and re-appear as another form of energy, which normally takes the form of sensible heat. In the atmosphere there is a very close relationship between potential energy and sensible heat, since as air parcels sink, their potential energy is converted into sensible heat and as they rise, their sensible heat is converted back into potential energy.

The equation for the total energy content of a unit mass of air can be used

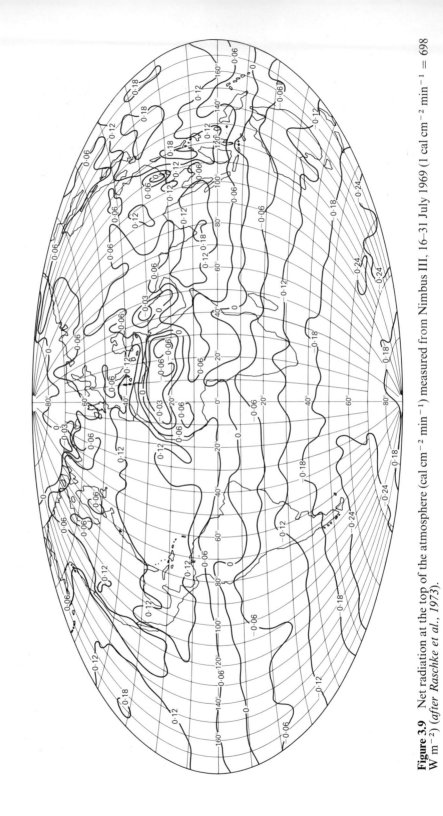

Figure 3.9 Net radiation at the top of the atmosphere (cal cm^{-2} min^{-1}) measured from Nimbus III, 16–31 July 1969 (1 cal cm^{-2} min^{-1} = 698 W m^{-2}) (*after Raschke et al., 1973*).

Plate 2 Gemini XI earth-sky view taken on 14 September 1966, looking northeast and showing parts of Libya, Chad, Niger and Algeria. The Mediterranean Sea is in the background. The desert sky is cloudless and the sandy areas with a high albedo are a light shade. *Reproduced by permission of the National Aeronautics and Space Administration.*

to study the energy balance of the tropical atmosphere. Outside of southern Asia, the mean north–south circulation of the tropical atmosphere can be considered as taking the form of two simple cells, with rising air near the equator and sinking air over the subtropical deserts. The low-level circulation of these cells form the north-east and south-east trade winds. Sinking air in the subtropics increases its temperature according to the dry adiabatic lapse rate, thus resulting in clear skies and low relative humidities. Subtropical deserts are largely a result of atmospheric subsidence leading to cloudless and rainless conditions. Large areas of the subtropics consist of ocean, and the clear skies result in a plentiful supply of solar radiation reaching the surface where it is mostly used to evaporate sea water. Water vapour evaporated over the subtropical oceans is mixed through the lower layers of the atmosphere by turbulence and convection and carried towards the equator by the trade winds. Near the equator, the trade winds enter the equatorial trough and here ascent takes place in localized weather systems and in particular in thunderstorms. In the thunderstorms, the latent heat released by the condensing water vapour is converted into sensible heat which in turn is transformed into potential energy by the rising air mass. In this way, the total potential energy (sensible heat + potential energy) of the rising air in the thundercloud is increased by the release of latent heat, and is then exported at high levels in the atmosphere into the subtropics and also into middle latitudes. In this manner the radiation deficit over the deserts is made up, as shown in Figure 3.10.

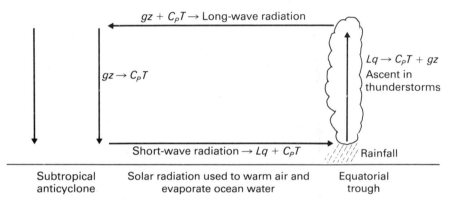

Figure 3.10 Energy conversions in a simple circulation cell (Hadley cell) (*after Lockwood, 1976*).

In a desert region there is a continual conversion of net radiation into sensible heat (C_pT). This contrasts with a vegetated region where there is a conversion of net radiation into both sensible heat and latent heat (Lq). Thus the air may be cooler in a vegetated area, but its total energy content ($C_pT + Lq + gz$) may be similar to or greater than that in a desert area. Now the important factor controlling the amount of absorbed short-wave radiation and therefore the net energy available for sensible heat transfer and evaporation is the albedo. The total energy ($C_pT + Lq$) imparted to the atmosphere is increased in the lower layers when the albedo is decreased. The albedo of deserts is considerably greater than the albedo of vegetated surfaces, so the actual net energy imparted to the atmosphere over deserts is less than that imparted over vegetated areas. Charney *et al.* (1976) have used this fact to suggest that convective rainfall will be decreased by a vegetated area being turned into a desert by drought or overgrazing. The important quantity determining convective precipitation is the negative gradient of total energy content with altitude and this negative gradient is increased when the albedo is decreased. Over low albedo, vegetated surfaces, evaporated water vapour is soon converted into sensible heat by rainfall, thus warming the atmosphere. Therefore at the desert boundaries the moist air over the vegetated surfaces will be warmer aloft than drier air over the high albedo desert surfaces. The desert air will therefore tend to sink at high levels relative to the air over the vegetated surfaces even though temperatures are very high at the desert surface. According to Charney *et al.* (1976) an extension towards the equator of desert conditions will cause an extension of the associated sinking air which will in turn tend to intensify the change towards desert conditions.

Charney *et al.* have applied their ideas to the Sahel region of West Africa and concluded that an increase of albedo in the Sahel alone would produce an appreciable decrease of precipitation. Even when there is no evapotranspiration, the precipitation is somewhat decreased because the increased albedo reduces the northward reach of the moist monsoon air from the Atlantic. These results are shown in Figures 3.11 and 3.12, which compare the longitudinally-averaged changes in rainfall in July over Africa produced by a change of albedo in the Sahel strip from 14 to 35 per cent, equivalent to change from full vegetation cover to full desert. The area is at present semi-desert on the southern boundary of the Sahara. Figure 3.11 shows the latitudinal distribution of rainfall in which the ground wetness at each point was kept fixed at its climatological value. In this model the evapotranspiration of the semi-arid zone was found by Charney *et al.* to be excessive. The results of carrying out a computation with essentially no evapotranspiration are shown in Figure 3.12. Their results do indicate a decrease of rainfall (shown by arrows in diagrams) immediately south of the latitude

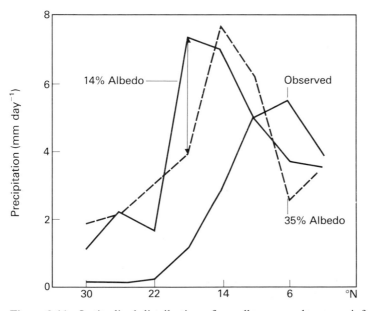

Figure 3.11 Latitudinal distribution of zonally averaged mean rainfall during July in North Africa for the case of fixed ground wetness and excessive evaporation (*after Charney et al., 1976*).

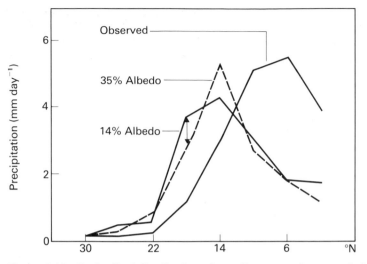

Figure 3.12   Latitudinal distribution of zonally averaged mean rainfall during July in North Africa for the case of variable ground wetness and negligible evaporation (*after Charney et al., 1976*).

at which an increase of rainfall causes an increase of vegetation and are in agreement with the observation that the recent drought in the Sahel was accompanied by increased rainfall further to the south. The ideas of Charney *et al.* are interesting because the 1968–73 drought has led to marked changes in the albedo of the Sahel which could in turn influence the rainfall. A similar feedback mechanism could operate in other tropical areas where the precipitation is largely convective and the surface winds are relatively light.

2.4 Ice surfaces

Snow is of great climatological importance because of its high short-wave albedo, about 70 per cent, as compared with other natural surfaces. Since it also has a high infrared emissivity, snow acts as a very effective cooling agent. In particular it is very effective in providing a positive feedback mechanism for strengthening seasonal changes. For example, Bauer and Dutton (1960) comment that there is a relationship between temperature at Madison, Wisconsin, and the extent of local snow cover. Thus it is observed that the cold weather of winter does not persist until the local area and that to the north have acquired a significant snow cover. It is also found that a persistent snow cover with its associated high albedo can appear quite rapidly and once established persist for some time even though there are many periods of sunshine to provide radiant energy at the surface. Similarly in the spring while the snow cover is on the ground, there is little modification of air masses over Wisconsin, since only about 30 to 40 per cent of the incoming short-wave radiation is absorbed by the surface. This effect continues until a warm air mass from a region free of snow moves over Wisconsin and removes the snow cover. Once the snow cover is removed,

Plate 3 Snow in the eastern USA. The white highly reflective surface of snow is clearly seen in this photograph. *Photographed by John Colwell from Grant Heilman, Lititz, PA.*

approximately 85 per cent of the incoming light is absorbed by the surface and much of this is available to warm the air over it, causing a rapid rise in average daily temperature at Madison.

The changes in albedo resulting from the melting of snow-covered ice are shown very clearly in Figure 3.13. This figure refers to the break-up of ice on a small lake in Wisconsin.

Even more interesting is the influence of ice and snow on world-wide climates. Measurements by meteorological satellites have shown that the albedo of the earth–atmosphere system in ice-covered areas is much greater than the albedo in ice-free regions. Therefore any increase in the area of ice and snow will increase the global albedo and lower world mean temperatures. Budyko (1969) and Sellers (1969) have constructed simple, one-dimensional, zonal mean models of the atmosphere in which the influence of snow cover upon the albedo of the earth's surface is taken into consideration. The effects of the poleward transport of heat due to atmospheric and oceanic circulations are incorporated in highly parameterized forms. They found that the positive feedback mechanism between the surface temperature and snow cover increases the sensitivity of their climatic models to changes in solar radiation. For example, in their models a small decrease in solar radiation is sufficient for the development of an extensive ice-sheet.

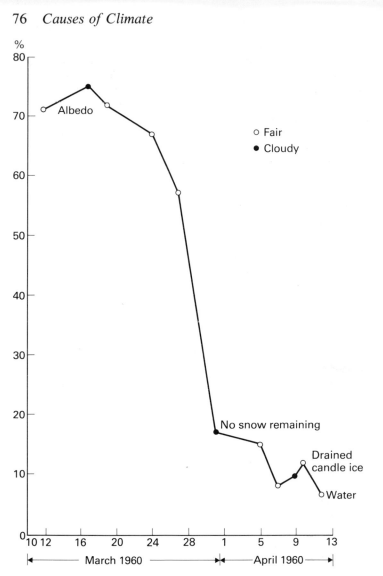

Figure 3.13   Albedo of lake-ice on small lake in Wisconsin, USA, during break-up and melting (*after Bauer and Dutton, 1960*).

Thus Budyko found with his model that a 1 per cent change in the solar constant with a fixed global albedo lowered mean global temperatures by 1·2 to 1·5 degree C. In contrast, a 1 per cent change in radiation with a corresponding changing albedo lowered the mean temperature by 5 degree C, while a 1·5 per cent change in radiation lowered it by 9 degree C. These reductions of temperature are accompanied by a southward advance of glaciation by 8–18 degrees of latitude, a distance corresponding roughly to the advance of the Quaternary glaciation. When solar radiation is reduced by 1·6 per cent, the ice-sheet reaches a mean latitude, about 50° N, after which it begins to advance southward all the way to the equator as a result of autodevelopment. This is associated with a sharp decline of global temperature, down to several tens of degrees below zero. Budyko found that if the ice-sheet was to occupy the entire surface from pole to the critical latitude of about 50° N, it would continue to advance toward the equator at the same lowered radiation value and, having reached still lower latitudes, it would continue to advance southward at the now existing magnitudes of radiation, and even at values exceeding the contemporary level.

Wetherald and Manabe (1975), using a complex mathematical model of the general circulation, have come to a similar conclusion. They found that once the earth was in a completely ice- or snow-covered condition, it would never recover from this state even with solar radiation at the present level. It must therefore be concluded that the earth has never been completely glaciated during its past history since there would have been little prospect of its ever having recovered from such a state.

3 Simple climatic models

It will now have become clear that a number of non-dimensional basic parameters exist which are of some importance to theoretical climatology. These concern the radiation, heat and water balances of the surface and atmosphere, and are listed below:

(a) Radiation
surface albedo (short-wave), α
surface emissivity, ε;
Ångström ratio, Å (quotient of effective long-wave radiation to that emitted at ground level).

(b) Heat Balance
Bowen ratio, β;
radiational index of dryness, R_N/LP (where P is precipitation);
Kelvin temperature at or near the ground surface, T.

(c) Water Balance
run-off ratio, f/P (where f is the run-off);
Exchangeable moisture contained in a vertical column of soil, m.

Relationships between the various non-dimensional parameters may be obtained as follows. For a land surface with a stable climate it follows that the multi-annual means of the soil storage terms of both heat and moisture must vanish. Using annual means (shown by a bar), it is possible to write for the land surface;

$$\bar{R}_N = \bar{C} + L\bar{E}_t$$

Since

$$\bar{\beta} = \bar{C}/L\bar{E}_t,$$

then

$$R_N = L\bar{E}_t(1 + \bar{\beta})$$

Also

$$\bar{P} = \bar{f} + \bar{E}_t$$

Combining the last two equations, it is possible to obtain

$$\frac{\bar{R}_N}{L\bar{P}} = (1 + \bar{\beta})\left(1 - \frac{\bar{f}}{\bar{P}}\right)$$

It also follows that

$$\frac{\bar{E}}{\bar{P}} + \frac{\bar{f}}{\bar{P}} = 1$$

Budyko (1956), supported by the analysis of actual precipitation, run-off, and net radiation data from a considerable number of catchments, further suggested that

$$\frac{\bar{f}}{\bar{P}} = 1 - \tanh\frac{\bar{R}_N}{L\bar{P}}$$

The above sets of equations contain two inputs, net radiation and precipitation, which determine the values of the outputs. This suggests that surface climates are basically a result of the radiation and precipitation inputs, and that the parameter $\bar{R}_N/L\bar{P}$ determines the relative values of the heat and water balances. Thus values of $\bar{R}_N/L\bar{P}$ up to 0·33 are typical of the tundra, values between 0·33 and 1 of the forest zones, from 1 to 2 of the steppes, and above 2 are typical of the semi-deserts and deserts.

The above equations can also be used to explore the range of climatic variation possible at a given climatological site. For a given site the range of variation of $\bar{R}_N$ tends to be rather limited. Run-off f depends in middle latitudes very much on the soil moisture state and tends to be high in winter and low in summer. Run-off is therefore very much a function of winter precipitation. The range of values of the Bowen ratio $\bar{\beta}$ gives rise to some difficulty. Both evaporation and sensible heat transfer are at a maximum in summer, so $\bar{\beta}$ represents some sort of average value of the minimum values of β experienced during the summer. The annual precipitation $\bar{P}$ shows by far the greatest variation, and therefore determines the variations in run-off. Indeed, since the variations in $\bar{P}$ are so large at a given site, they probably determine the local variations in climate.

The general importance of the other radiation, heat and water balance parameters has already been illustrated in earlier sections of this chapter. The surface albedo determines the amount of energy absorbed and thus the total sum of the energy balance of the surface for a given radiation input.

If the albedo is high, the amount of energy at the surface available for distribution is small.

Surface temperature is important because together with the surface emissivity it determines the rate of infrared radiation loss. Thus infrared losses are very high in regions of high surface temperature, such as the subtropical deserts, and relatively low from ice-sheets. All surface infrared losses are complicated by the large and continuous atmospheric counter radiation. Surface temperature is also important in that it partly controls the Bowen ratio. Thus at high temperatures most of the net available energy goes towards evaporation.

The soil moisture state is important in that it determines both the run-off and evaporation rates, and is in turn determined by the precipitation. If the soil becomes dry, the evaporation rate is limited and the Bowen ratio rises above the level appropriate for the particular surface temperature.

References

BAUER, K. G. and DUTTON, J. A. 1960: *Flight investigations of surface albedo.* Technical Report **2**, Department of Meteorology, University of Wisconsin.

BUDYKO, M. I. 1956: *The heat balance of the earth's surface.* Translated by N. I. Stepanova. Washington: US Weather Bureau.

1969: The effect of solar radiation variations on the climate of the earth. *Tellus* **21**, 611–19.

CHARNEY, J. G. 1975: Dynamics of deserts and droughts in the Sahel. *Quarterly Journal Royal Meteorological Society* **101**, 193–202.

CHARNEY, J. G., STONE, P. H. and QUIRK, W. J. 1975: Drought in the Sahara: a biogeographical feedback mechanism. *Science* **187**, 434–5.

1976: Drought in the Sahara: insufficient biogeographical feedback? *Science* **191**, 100–2.

GRINDLEY, J. 1972: Estimation and mapping of evaporation. In IASH, *World water balance* **1**, 200–213. Gentbrugge.

JARVIS, P. G., JAMES, G. B. and LANDSBERG, J. J. 1976: Coniferous forest. In MONTEITH, J. L., editor, *Vegetation and the atmosphere* **2**. London: Academic Press, 171–240.

LINACRE, E. T. 1967: Further notes on a feature of leaf and air temperatures. *Archiv Meteorologie Geophysik und Bioklimatologie*, Ser B **15**, 422–36.

LOCKWOOD, J. G. 1976: *The physical geography of the tropics: an introduction.* Kuala Lumpur: Oxford University Press.

MONTEITH, J. L. 1965: Evaporation and environment. In *The State and movement of water in living organisms*, 205–34. 19th Symposium Society Experimental Biology.

1975: *Vegetation and the atmosphere* **1**. London: Academic Press.

1976: *Vegetation and the atmosphere* **2**. London: Academic Press.

MONTEITH, J. L. and SZEICZ, G. 1961: The radiation balance of bare soil and vegetation. *Quarterly Journal Royal Meteorological Society* **87**, 159–70.

NAMIAS, J. 1969: Seasonal interactions between the North Pacific Ocean and the atmosphere during the 1960's. *Monthly Weather Review* **97**, 173–92.

1970: Macroscale variations in sea-surface temperatures in the North Pacific. *Journal Geophysical Research* **75**, 565–82.

PENMAN, H. L. 1948: National evaporation from open water, bare soil and grass. *Proceedings Royal Society*, Series A **193**, 120–45.

PRIESTLEY, C. H. B. 1966: The limitation of temperature by evaporation in hot climates. *Agricultural Meteorology* **3**, 241–6.

PRIESTLEY, C. H. B. and TAYLOR, R. J. 1972: On the assessment of surface heat flux and evaporation using large-scale parameters. *Monthly Weather Review* **100**, 81–92.

RASCHKE, E. 1973: *The radiation balance of the earth–atmosphere system from Nimbus 3 radiation measurements.* NASA Technical Note D-7249.

RIPLEY, E. A. and REDMANN, R. E. 1976: Grassland. In MONTEITH, J. L., editor, *Vegetation and the atmosphere* **2**, 351–98. London: Academic Press.

ROLL, H. U. 1965: *Physics of the marine atmosphere.* New York: Academic Press.

RUTTER, A. J. 1975: The hydrological cycle in vegetation. In MONTEITH, J. L., editor, *Vegetation and the atmosphere* **1**, 111–54. London: Academic Press.

SELLERS, W. D. 1969: A global climatic model based on the energy balance of the earth–atmosphere system. *Journal Applied Meteorology* **8**, 392–400.

STEWART, J. B. and THOM, A. S. 1973: Energy budgets in pine forest. *Quarterly Journal Royal Meteorological Society* **99**, 154–70.

WETHERALD, R. T. and MANABE, S. 1975: The effects of changing the solar constant on the climate of a general circulation model. *Journal Atmospheric Sciences* **32**, 2044–59.

4
The Distribution of Surface Climates: Atmospheric Circulation Patterns

1 Controlling factors of the general circulation

The planet earth receives heat from the sun in the form of short-wave radiation, but it also radiates an equal amount of heat to space in the form of long-wave radiation. This balance of heat gained equalling heat lost only applies to the planet as a whole over several annual periods; it does not apply to any specific area for a short period of time. The equatorial region absorbs more heat than it loses, while the polar regions radiate more heat than they receive. Nevertheless, the equatorial belt does not become warmer during the year, nor do the poles become colder, because heat flows from the warm to the cold regions, thus maintaining the observed temperatures. An exchange of heat is brought about by the motion of the atmosphere and the upper layers of the oceans, thus forming the general circulation of the atmosphere and the oceans. The general circulation would probably assume a very simple form in the absence of the modifying influence of the earth's rotation, consisting of two simple direct cells, one in each hemisphere. It is therefore possible to state that the general atmospheric circulation is 'driven' by inequalities in radiation distribution and 'shaped' by the earth's rotation.

2 Distribution of radiation

Figure 4.1 shows a generalized model of the annual radiation budget of the global atmosphere. The assumption is made that 100 units of radiant energy are received at the top of the atmosphere, and the model then traces the various interactions that the radiation undergoes in the atmosphere. It is interesting to note that nearly 50 per cent of the incoming radiation penetrates through the atmosphere and is absorbed by the earth's surface. The atmosphere is therefore semi-transparent to the incoming short-wave radiation but in contrast is almost opaque to infrared which is absorbed strongly by water vapour and carbon dioxide. The geographical distributions of the various radiation exchanges shown in Figue 4.1 are now considered in more detail.

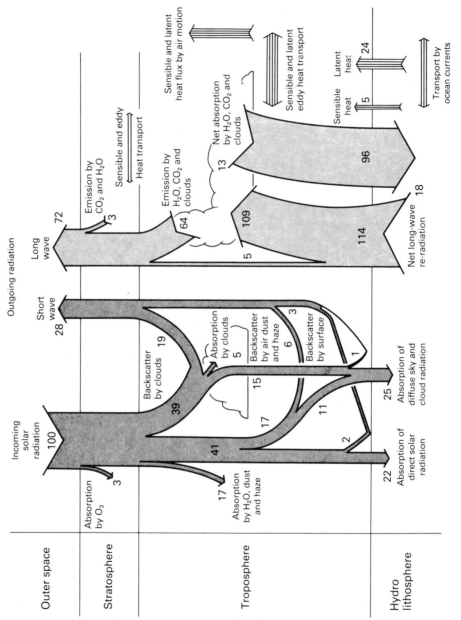

Figure 4.1 Schematic diagram showing the interactions that radiation undergoes in the atmosphere (*after Schneider and Mesirow, 1976, and Rotty and Mitchell, 1974*).

2.1 Global radiation

Global radiation is the sum of all short-wave radiation received both directly from the sun and indirectly from the sky, on a horizontal surface. Generalized isolines of average annual global radiation are shown in Figure 4.2, which conveys a very general picture of the distribution of global radiation. It looks deceptively simple because the original map was based on few observations which, particularly over the oceans, were restricted to isolated island stations. The actual distribution of global radiation reflects closely astronomical factors and the distribution of cloud. Thus the areas receiving most global radiation are found in the subtropics where there are unusually clear skies because of the prevailing anticyclonic conditions.

Hanson (1976) has attempted to map tropical distributions of global radiation using two years of satellite-derived cloud cover data and some surface radiation measurements. The results are shown in Figure 4.3 and there are a number of features in the distribution which result directly from variations in cloudiness. For example, at 15° N the radiation reaches a maximum in April (670 cal cm^{-2} day^{-1}), apparently associated with the subtropical high-pressure zone, and this maximum progresses northward to near 30° N by June and July. There is also a maximum in solar radiation in a zone within a few degrees of the equator during nearly all of the year, but adjacent to this equatorial maximum are minima in the zonal radiation field. In the northern hemisphere the minimum is a result of cloudiness in the intertropical convergence zone, which is located at $2\frac{1}{2}$° to 5° N in March but moves to 5° to 10° N by September and October. This minimum is present during the entire year, and is most intense during July when the radiation at 5° to $7\frac{1}{2}$° N is only 410 cal^{-2} day^{-1}. In the southern hemisphere a similar but less developed radiation minimum exists from September till April.

Hanson suggests that the global average solar radiation at the earth's surface is 0·286 cal cm^{-2} min^{-1}. If a solar constant of 1·95 cal cm^{-2} min^{-1} is assumed and the earth intercepts one fourth of this amount, or 0·488 cal cm^{-2} min^{-1}, then this means that 58·6 per cent of the extraterrestrial radiation intercepted by the earth is incident at the earth's surface.

2.2 Albedo

Radiation reflected directly back to space from the earth constitutes a loss of available energy to the earth–atmosphere system. The distribution of albedo values over the earth's surface must therefore be considered together with the global radiation. The annual albedo of the earth–atmosphere system together with the solar radiation absorbed annually in the earth–atmosphere system are shown in Figures 4.4 and 4.5. Both maps are based on meteorological satellite observations and the absorbed solar radiation is partly a combination of Figures 4.4 and 4.2. The albedo map clearly reveals the land–sea distribution and the general atmospheric circulation as it is represented by the mean cloud patterns over both hemispheres. The high-reaching convective clouds associated with the intertropical convergence

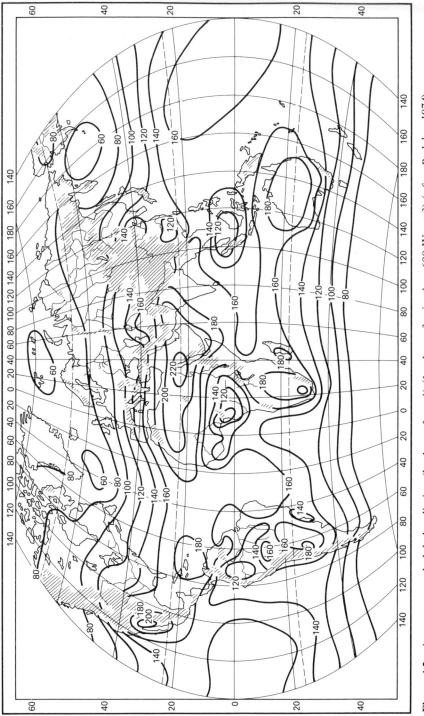

Figure 4.2 Average annual global radiation (kcal cm^{-2} yr^{-1}) (1 cal cm^{-2} min^{-1} = 698 W m^{-2}) (*after Budyko, 1974*).

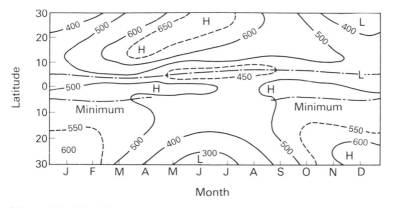

Figure 4.3 Monthly average zonal solar irradiance at the earth's surface (cal cm⁻²
day⁻¹) based on 2 years of satellite data from February 1965 to January 1967 (1 cal
$cm^{-2}\ min^{-1} = 698\ W\ m^{-2}$) *(after Hanson, 1976)*.

zone and partly with the Asian monsoon appear as a belt of relatively high
albedos of more than 25–30 per cent. Similarly low persistent stratus clouds
along the western coastal areas of North and South America and Africa
appear with albedos between 25 and 35 per cent. The albedo of both polar
regions is considerably higher than 50 per cent because of the associated
permanent snow and ice-fields. Regions of major gain of radiative energy
are the oceanic areas in the subtropics of both hemispheres. The planetary
albedo shown in Figure 4.4 is made up of three main components. These
are the light reflected from the actual land and sea surfaces of the earth,
the light reflected by clouds, and the light scattered upwards by the
atmosphere. Estimates of the albedo of the ground and also locations of
persistent cloud fields and of ice and snow can be made when travelling or
otherwise changing cloud-fields are removed by displaying only the lowest
observed satellite albedo value in each area. This approach is based on the
simple assumption that the albedo of the earth–atmosphere system is higher
over each area in the presence of clouds than for a cloud-free atmosphere.
A map produced by Raschke *et al.* (1973) in this manner is shown in
Figure 4.6. It clearly shows the much lower albedos observed over the oceans
as compared with those observed over the continents. Similarly over Africa
the high albedos of the desert surfaces are clearly observed, with a relatively
sharp boundary where the southern Sahara grades into regions with rather
more vegetation. Minimum albedos greater than 40 per cent belong to ice-
fields at their observed smallest extent during July over the Arctic and during
January over the Antarctic, respectively.

2.3 Outgoing long-wave radiation
The outgoing long-wave radiation, as observed by satellite, is shown in
Figure 4.7. The distribution shown reflects the temperatures of the emitting
surfaces. Thus high clouds are cold and have a low emission, while low-level
surfaces are warm and have a high rate of emission. Over Africa, low rates

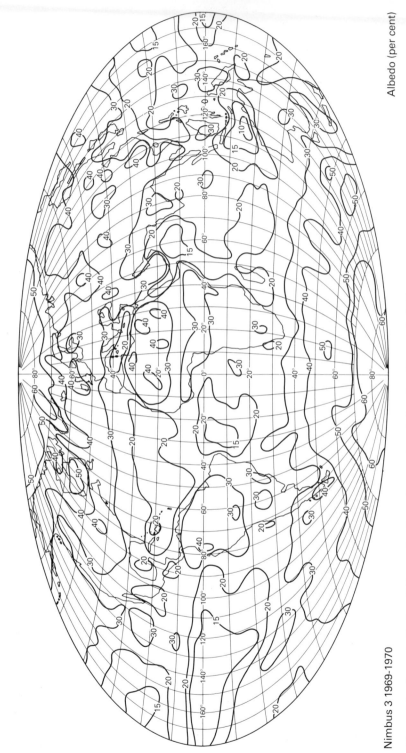

Nimbus 3 1969-1970

Albedo (per cent)

Figure 4.4 Annual albedo of the earth–atmosphere system (*after Raschke et al., 1973*).

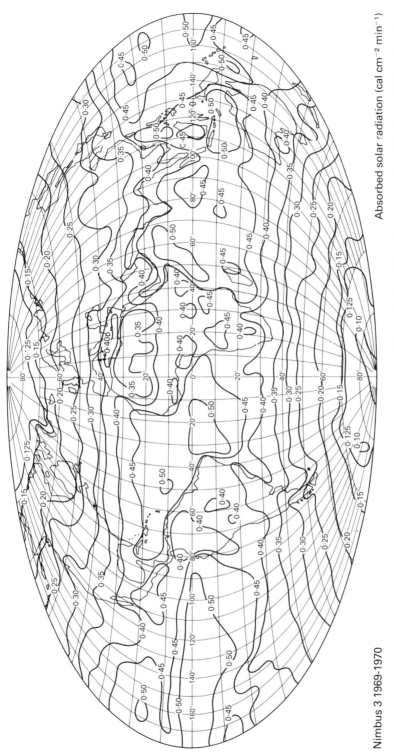

Nimbus 3 1969-1970

Absorbed solar radiation (cal cm^{-2} min^{-1})

Figure 4.5 Solar radiation absorbed annually in the earth–atmosphere system (cal cm^{-2} min^{-1}) (1 cal cm^{-2} min^{-1} = 693 W m^{-2}) (*after Raschke et al., 1973*).

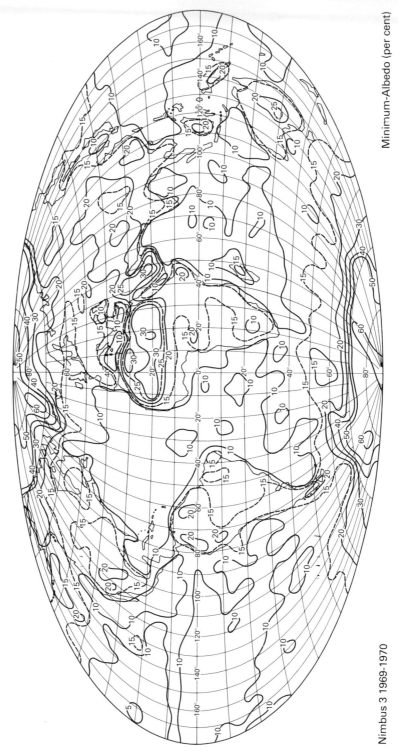

Nimbus 3 1969-1970

Minimum-Albedo (per cent)

Figure 4.6 Minimum albedo (*after Raschke et al., 1973*).

Nimbus 3 1969-1970

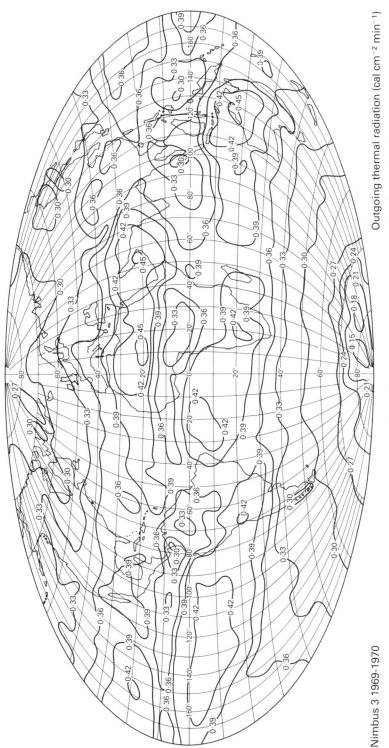

Outgoing thermal radiation (cal cm⁻² min⁻¹)

Figure 4.7 Long-wave radiation emitted annually to space (cal cm⁻² min⁻¹) (1 cal cm⁻² min⁻¹ = 698 W m⁻²) (*after Raschke et al., 1973*).

of emission are observed from the high convective clouds of the intertropical convergence zone while over the Sahara high rates are observed since the atmosphere is clear and the warm sands of the desert surface are visible.

Outgoing radiation from the whole globe corresponds to an effective temperature of 255 K, and the planetary albedo is found to be about 30 per cent.

Plate 4   A mosaic of pictures of the northern hemisphere taken by Essa 9 on 22 September 1971. In general the bright cloudy areas indicate upward motion in the atmosphere, while extensive cloud-free areas are probably the result of widespread subsidence. Cloudy areas have the highest albedos, followed by the areas of snow and ice, and by desert sands. *Reproduced by permission of the National Oceanic and Atmospheric Administration.*

2.4 Radiation balance

Figure 4.8 shows the geographical distribution of annual net radiation at the earth's surface only, for the atmosphere is excluded. This figure reveals that the annual means of the net radiation balance over the greater part

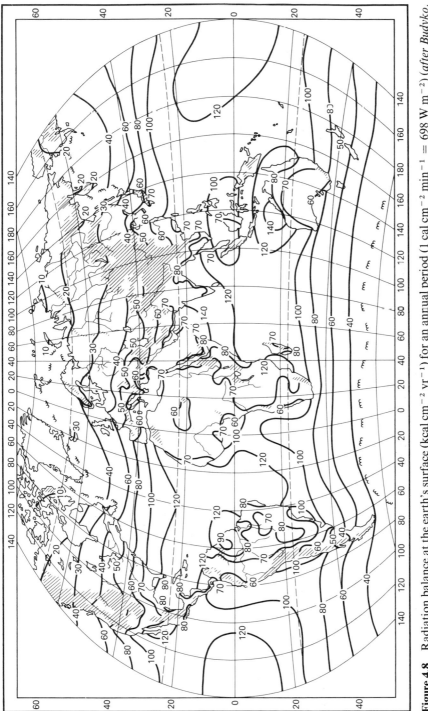

Figure 4.8 Radiation balance at the earth's surface (kcal cm^{-2} yr^{-1}) for an annual period (1 cal cm^{-2} min^{-1} = 698 W m^{-2}) (*after Budyko, 1974*).

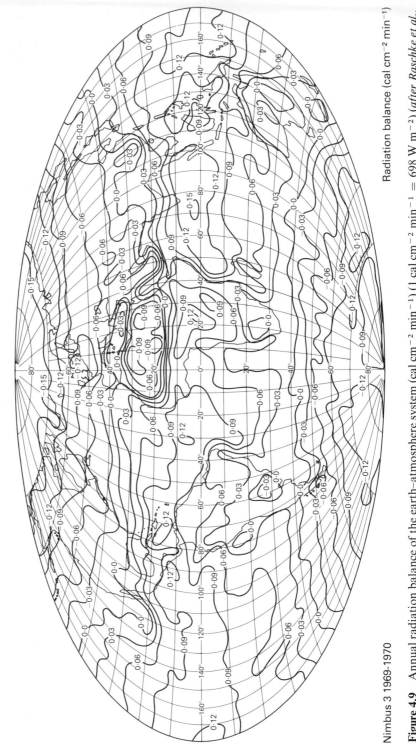

Nimbus 3 1969-1970

Radiation balance (cal cm^{-2} min^{-1})

Figure 4.9 Annual radiation balance of the earth–atmosphere system (cal cm^{-2} min^{-1}) (1 cal cm^{-2} min^{-1} = 698 W m^{-2}) (*after Raschke et al., 1973*).

of the earth's surface are positive, and thus signifies that the absorbed short-wave radiation is greater than the long-wave outgoing radiation. This pattern is the result of the greater transparency of the atmosphere for short-wave radiation in comparison with long-wave radiation, and the excess of energy at the earth's surface is transferred to the atmosphere by turbulent heat exchange and by evaporation.

A satellite estimate of the annual radiation balance of the earth–atmosphere system is shown in Figure 4.9. Positive values, mostly found between 40° N and 40° S, imply a heat gain by the surface–atmosphere system, while negative values elsewhere imply a general heat loss to space. The net radiation integrated over the whole globe must come to zero, because there is, over an annual period, almost an exact balance between solar energy absorbed and infrared radiation to space. Significant changes in the earth's radiation budget occur within latitudinal zones, especially in the tropics. Thus the areas of the major gains of radiative energy are the oceans in the subtropics of both hemispheres, while the African and Arabian deserts at the same latitude actually have a radiative deficit.

Seasonal changes of radiation balance (Figure 4.10) are dominated by seasonal variations in solar declination. The radiative balance of the surface–atmosphere system is positive during the whole year only in the narrow equatorial zone between the latitudes 10° N to 10° S, for elsewhere the sign

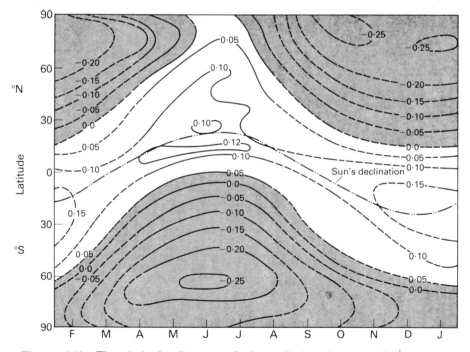

Figure 4.10 Time–latitude diagram of the radiation balance of the earth–atmosphere system (cal cm^{-2} min^{-1}) (1 cal cm^{-2} min^{-1} = 698 W m^{-2}) (*after Raschke et al., 1973*).

of the radiation balance changes twice a year. For about three summer months in a year the radiation balance of the whole of each hemisphere is positive, but in late summer, zones of negative balance arise near the poles and then gradually spread toward the equator, reaching latitude thirty after five months; a similar process of retreat begins in the spring.

3 Atmospheric motions

Two forces control either directly or indirectly most of the motions found in the atmosphere. Gravity is finally responsible for generating most motions and accelerations, while friction is largely responsible for keeping objects at rest or reducing velocities. The difference between the forces is that gravity accelerates objects vertically downwards towards the centre of the earth, while friction is a mechanical force of resistance which acts when there is relative motion between two bodies in contact.

Gravity can be responsible for producing motions within the atmosphere. The pressure intensity in the atmosphere at points on a level plane is less than at points at a lower level, since there is less of the atmosphere above the former. This decrease of pressure (Δp) with increasing altitude (Δz) may be expressed in the form:

$$\frac{\Delta p}{\Delta z} = -g\rho$$

where $\Delta p/\Delta z$ is the change of pressure with altitude, g is the downward acceleration due to gravity, and ρ is the density of the air.

In the undisturbed atmosphere, pressure will be equal in all horizontal surfaces, since no unbalanced forces are acting within the atmosphere. Forces within the atmosphere must be the same in all directions, otherwise the atmosphere will flow until the forces are equalized. If pressure varies in a horizontal plane (note that it always varies in the vertical), the forces in the horizontal plane will not balance and fluid will flow from high to low pressure. This movement is due to the horizontal pressure gradient, the original spot pressures being created by the force of gravity. Variations of pressure within a horizontal plane could be induced by heat flowing into one point of the atmosphere and so generating a rise in temperature. Density varies with temperature, so a rise in temperature will cause a decrease in density and thus a fall in pressure. Many motions within both the atmosphere and oceans are generated by differential heating which results in unequal horizontal pressure distributions, the fluid then flowing under the influence of gravity.

Although there is a long-term balance between incoming and outgoing radiation, considerable imbalances exist both locally and seasonally, and it has already been shown that there is a substantial excess of net radiation in low latitudes and a deficit towards the poles. Alternatively, it can be considered that the mean atmospheric temperature in equatorial latitudes is lower, and in polar latitudes higher, than those appropriate to the local radiative balance, and that this situation is possible because of the global-scale mixing performed by the general atmospheric circulation. The atmospheric circulation is maintained against the various frictional forces which tend to rapidly destroy it by the heat energy from the sun, and the circulation would assume a very simple form but for the modifying influence of the earth's rotation. This simple circulation, in the absence of the earth's rotation, would probably consist of two simple cells with rising air over the equator and sinking air over the two poles.

3.1 The earth's rotation

Many of the large-scale motions upon the earth's surface are influenced by the rotation of the earth. This influence can be illustrated by exploring the motion of a particle on a flat rotating disc, which is shown in Figure 4.11.

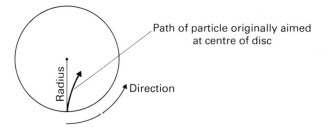

Figure 4.11 Motion on a rotating disc. Projectile moving towards the centre of the rotating disc appears to be deflected towards the right.

Since the disc is formed of a rigid solid, all points on the disc rotate around the axis in the same time interval. Now some points are further away from the central axis than others and therefore have to travel a greater distance during the period of rotation than points nearer to the axis. Because of the increasing distance to be travelled at greater distances from the axis, it follows that the actual speed of movement will increase with increasing remoteness from the axis. Thus the actual speed of movement will be greatest for points on the rim of the disc and least for points near the centre.

Consider a particle which is situated on the edge of a non-rotating disc. According to Newton's first law, an object not at rest and not subject to external forces will move in a straight line at a uniform speed. So in the absence of friction, a particle projected from the rim towards the central axis of the disc will move along a straight line which will form one of the radii of the disc.

Imagine that the disc is rotating and that the particle moves across the disc in the absence of friction. At the starting point on the rim, the particle is at rest relative to the rotating disc and it will have a velocity which is appropriate for its distance from the axis. If the particle is now projected towards the centre of the disc, its velocity will be made up of two components—the component towards the centre and a component at a right angle to the first due to the rotation of the disc. As the particle moves towards the centre it passes over parts of the disc where the speed of motion is less than that on the rim, so the particle, which retains its original speed, will appear to move sideways relative to the disc. If the disc rotates in an anticlockwise direction, the particle will appear to be deflected towards the right of the radius line from its starting position on the rim.

The earth's rotation about its axis at any point can be divided into three components (Figure 4.12), the most important of which is the one in a horizontal plane about an axis vertically upwards. At the poles the horizontal

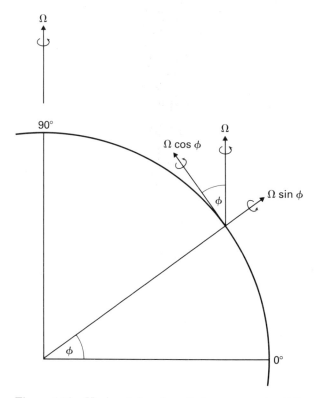

Figure 4.12   Horizontal and vertical components of the earth's rotation about its axis: Ω is the rate of the earth's rotation; φ is the latitude.

component (about a vertical axis) will be equal to the rate of the earth's rotation, while away from the pole it will vary with sine (latitude), becoming zero at the equator. The horizontal component of the earth's rotation is such that the earth simulates a flat disc rotating anticlockwise in the northern hemisphere and clockwise in the southern. From the previous discussion it follows that particles moving horizontally outwards from the centre of the disc appear to an observer stationed there, to be deflected to the right in the northern hemisphere and to the left in the southern. To the observer on the rotating earth it appears that a force is acting on the particles which causes them to be deflected, and this apparent force per unit mass is termed the 'Coriolis force' or 'deviating force' or 'geostrophic force'. The Coriolis force is strictly a three-dimensional vector which is everywhere at right angles to both the earth's axis (in the plane of the equator) and the velocity of the object in motion, but normally only the horizontal component is of interest. The acceleration produced by the horizontal component of the Coriolis force is $2\Omega V \sin \varphi$, where Ω is the rate of the earth's rotation, V is the velocity of the particle and φ is the latitude. This acceleration is always small, having a magnitude of $1.5 \times 10^{-4} \times$ velocity cm s^{-2} at the poles and decreasing to zero at the equator ($\sin (0°) = 0$). The quantity $2\Omega \sin \varphi$ is known as the Coriolis parameter.

On the laboratory scale the Coriolis effect is masked completely by frictional forces, and therefore need not be considered. Similar considerations apply to small-scale movements in the natural environment, where friction effects are large and therefore rivers are not usually influenced by the Coriolis effect. Both the frictional and driving forces observed in large-scale motions in the atmosphere are small and are of the same order of magnitude as, or smaller than, the Coriolis effect, and so in both these cases the Coriolis effect is of some importance. The motion will be dominated by whatever force has the greatest magnitude, and it is only in very large-scale motion systems in the atmosphere and oceans that the other forces are small enough to allow the Coriolis force to become dominant.

The influence of the earth's surface on air flow is often detectable up to an altitude of 2 km, and this layer is known as the planetary boundary layer. At the top of the planetary boundary layer the wind can to a good approximation be considered to be equal to the geostrophic wind, which is an imaginary wind blowing parallel to the isobars at a speed which is inversely proportional to the spacing of the isobars. On a non-rotating earth the winds would blow directly down the pressure gradient, but on a rotating earth (Figure 4.13) the wind velocity assumes a magnitude so that the

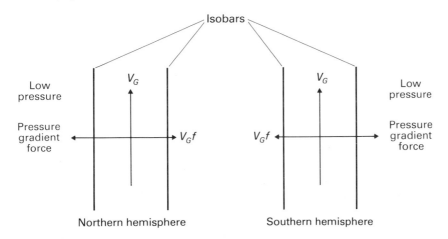

Figure 4.13 The geostrophic wind: V_G is the geostrophic wind; f is the horizontal component of the Coriolis force. The pressure gradient force acts from high to low pressure.

horizontal component of the Coriolis force balances the pressure gradient force. That is:

$$\text{horizontal pressure gradient force} = V_{G}f\rho$$

where V_{G} is the geostrophic wind; f is the horizontal component of the Coriolis force ($2\Omega \sin \varphi$); and ρ is the air density.

Thus low pressure is to the left of the wind arrow in the northern hemisphere, and to the right in the southern. Below the top of the planetary boundary layer the wind is increasingly deflected towards low pressure, until near the surface there is marked cross-isobaric flow and the wind speed is considerably less than the geostrophic value.

Any small region of the atmosphere contains two components of rotation about a vertical axis. The first is the horizontal component of the earth's rotation and the second is due to local circulations within the atmosphere. Since the earth rotates towards the east, in an anticlockwise manner in the northern hemisphere, the first component is also anticlockwise in the northern hemisphere. In a low-pressure system there is general convergence of air at low levels towards the centre of the system, and the air imported contains the dominant anticlockwise spin due to the earth's rotation. The convergence into the system intensifies the latent rotation of the air and produces the cyclonic spin of the depression. In the southern hemisphere the horizontal component of the earth's rotation acts in the opposite sense, so clockwise rotation occurs in low-pressure systems. General outflow takes place at the surface in anticyclones (high-pressure systems) and this leads to the opposite effect to that observed in depressions, that is clockwise rotation in the northern hemisphere and anticlockwise in the southern.

The rotation of the earth can affect the general circulation in a second way. The angular momentum per unit mass of a body rotating about a fixed axis is the product of the linear velocity of the body and the perpendicular distance of the body from the axis of rotation. Now in the upper atmosphere, in the absence of significant friction or other forces, angular momentum remains constant over periods of a few days. Rings of air flowing poleward at high levels from near the equator move nearer to the earth's axis, because of the spherical shape of the earth, and since the angular momentum is conserved, the eastward velocity of the air increases. Thus, on average, the eastward velocity of the air between about 6 km and 14 km increases away from the equator, reaching velocities of 30 to 60 m s^{-1} between 30° and 40° N and S, appearing on weather charts as the subtropical westerly jet streams. A jet stream is a fast narrow current of air, usually found in the upper troposphere, and it is generally some thousands of kilometres in length, hundred of kilometres in width and a few kilometres in depth. The subtropical jet streams mark the poleward limit of the outward flow from the equator, and there is extensive subsidence out of them. Therefore the two simple direct cells, which would be observed on a non-rotating earth, are confined by the earth's rotation to the region between 40° N and 40° S.

It should now be clear that the earth's rotation shapes the general circulation, for if the earth were to rotate in the opposite sense, the direction of rotation of weather systems would be changes as also would the direction of the major wind systems.

3.2 Atmospheric thermal patterns

The basic thermal pattern of the lower half of the atmosphere in the temperate latitudes is partly controlled by the prevailing mean surface temperatures. The mean thermal pattern may be usefully investigated by using the concept of thickness lines. The thickness (or the depth) of an isobaric layer is given by

$$z - z_0 = \frac{RT_m}{g} \ln\left(\frac{p_0}{p}\right)$$

where z and z_0 are the heights above sea-level of the top and bottom of the isobaric layer; p and p_0 are the pressures at the top and bottom of the isobaric layer; R is the gas constant for dry air; and T_m is the mean temperature of the layer. Clearly the thickness of an atmospheric layer bounded by two fixed pressure surfaces, 1,000 and 500 mbar for example, is directly proportional to the mean temperature of the layer. Thus low thickness values correspond to cold air and high thickness values to warm air.

The geostrophic wind velocity at 500 mbar is the vector sum of the 1,000 mbar geostrophic wind and the theoretical wind vector blowing parallel to the 1,000–500 mbar thickness lines, with a velocity proportional to their gradient. This theoretical wind is known as the thermal wind and is shown in Figure 4.14. The magnitude (V_T) of the thermal wind is given by the expression:

$$V_T = \left| \frac{g}{f} \frac{\Delta z'}{\Delta n} \right|$$

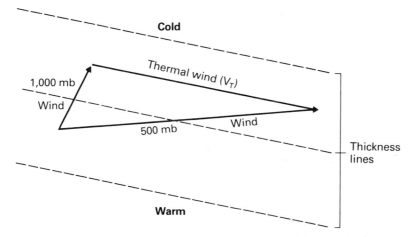

Figure 4.14 The thermal wind (V_T) is the vector difference between the 1,000 mbar wind and the 500 mbar wind.

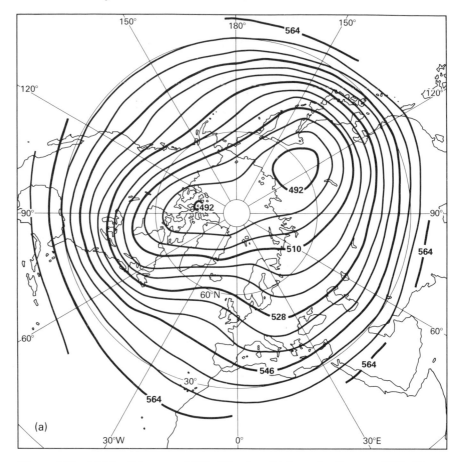

Figure 4.15 (a) Monthly mean 1,000–500 mbar thickness for January 1951–66.

where g is the gravitational acceleration; f is the Coriolis parameter; and $\Delta z'/\Delta n$ is the horizontal gradient of the thickness of the layer.

Inasmuch as the thickness is but another expression for the mean temperature of the isobaric layer, the thermal wind will blow along the mean isotherms with lower temperature to the left in the northern hemisphere. Thus in temperate latitudes the thermal wind will be westerly, and also since horizontal temperature gradients are relatively steep, it will be strong. The upper winds are the vector sum of a rather weak surface wind field and a vigorous thermal wind field. This implies that in the lower middle-latitude troposphere, winds will become increasingly westerly with altitude and that the upper wind field is strongly controlled by the thermal wind. A parallel therefore exists between the mean topography of the 500 mbar level and the mean 1,000–500 mbar thickness patterns, as shown in Figure 4.15.

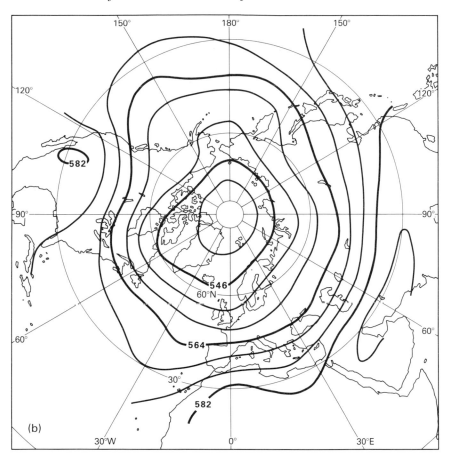

(b) Monthly mean 1,000–500 mbar thickness for July 1951–66.

The mean January thickness patterns (Figure 4.15) for the northern hemisphere show two dominant troughs near the eastern extremities of the two continental landmasses, while ridges lie over the eastern parts of the oceans. A third weak trough extends from north Siberia to the eastern Mediterranean. Climatologically, the positions of the main troughs may be associated with cold air over the winter landmasses, and the ridges with relatively warm sea surfaces.

In July, the mean ridge in the thickness pattern found in January over the Pacific has moved about 25° west, and now lies over the warm North American continent, while there is a definite trough over the east Pacific. Patterns elsewhere are less marked, but a weak trough does appear over Europe, and may perhaps be connected with the coolness of the North Sea, the Baltic, the Mediterranean and the Black Sea.

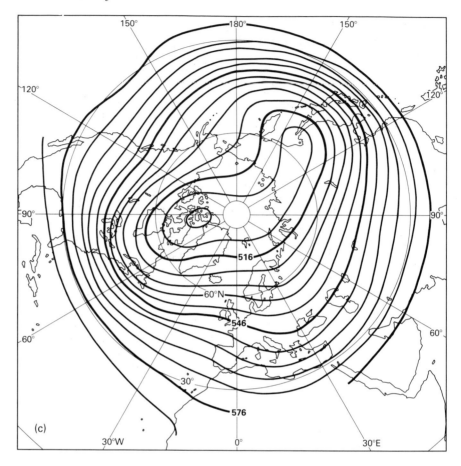

Figure 4.15   (c) Monthly mean 500 mbar contours for January 1951–66.

3.3 Atmospheric wind and pressure systems

The sun's radiation, through its absorption by the ground, causes a temperature gradient (and therefore a pressure gradient) between equator and pole,
and thereby drives the general atmospheric circulation. Under these conditions, if the earth did not rotate, a poleward flux of air would exist aloft
while a compensating return flow towards the equator would exist at low
levels. Such a simple circulation cell is often known as a Hadley cell, but
because of the rotation of the earth it does not extend from equator to pole
in the atmosphere, but instead it is confined to low latitudes.

A schematic representation of the mean meridional circulation in the
northern hemisphere during winter is shown in Figure 4.16. The simple
Hadley cell circulation is clearly seen south of 30° N. Eastward angular
momentum is transported from the equatorial latitudes to the middle latitudes

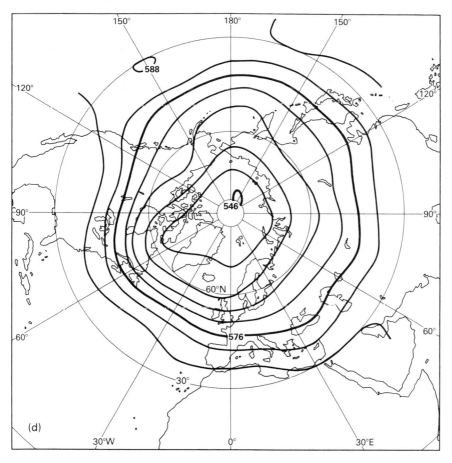

(d) Monthly mean 500 mbar contours for July 1951–66.
Isopleths are at intervals of 6 geopotential decametres (*after Moffitt and Ratcliffe, 1972*).

by nearly horizontal eddies, 1,000 km or more across, moving in the upper troposphere and lower stratosphere. Such a transport leads to an accumulation of eastward momentum between 30° and 40° latitude, where a strong meandering current of air, generally known as the subtropical westerly jet stream, develops. The cores of strong winds in both the hemispheres in both the seasons occur at an altitude of about 12 km, as shown in Figure 4.17.

More momentum than is necessary to maintain the subtropical jet streams against dissipation through internal friction is transported to these zones of strong winds. The excess is transported downwards to maintain the eastward flowing surface winds of the middle latitudes against the ground friction. The air subsiding from the jet streams also forms the subtropical anticyclones (Figure 4.18). The supply of eastward momentum to the earth's surface in middle latitudes tends to speed up the earth's rotation. Counteracting such

Plate 5   American geostationary satellite picture of the Atlantic Ocean showing the surrounding continents and cloud formations, taken on 27 September 1974. *Reproduced by permission of Météorologie Nationale, Paris.*

a continuous speeding up of the earth's rotation, air flows from the sub-tropical anticyclones towards the equatorial regions, forming the so called trade winds. The trade winds, with a strong component directed towards the west, retard the earth's rotation, and in turn pick up eastward momentum.

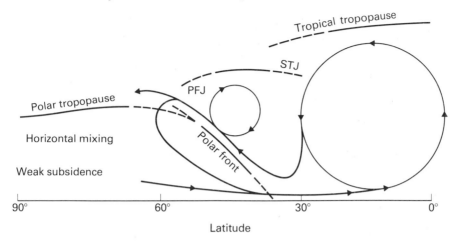

Figure 4.16 Schematic representation of the meridional circulation and associated jet stream cores in winter. STJ, subtropical jet stream; PFJ, polar front jet stream (*after Palmen, 1951*).

The trade winds also pick up moisture evaporated from the oceanic surfaces along their trajectories close to the earth's surface. As they ascend over the equatorial latitudes, the water vapour condenses, liberating the heat of condensation to the ascending air and giving rise to the high-rainfall regions of the equatorial world.

Poleward of latitudes 30° N and S, the meridional components of motion near the surface and in the upper levels are the reverse of those that occur in the lower latitudes. There is a marked poleward component in the winds near the surface and an equatorial component in the upper levels, with an average descending motion in the subtropics and air ascending in the higher latitudes. The poleward-moving air picks up moisture from the underlying ocean surface and transports it poleward; the condensation of the water vapour in the ascending branch releases latent heat and gives rise to the rainfall of the middle latitudes.

An idealized mean surface wind circulation with the associated pressure distribution, appropriate to a uniform earth, may be described as follows:

(a) *The equatorial trough.* A shallow belt of low pressure on or near the equator with light or variable winds.

(b) *The trade winds.* Between the equatorial trough and latitudes 30° N and 30° S are north-east winds in the northern hemisphere and south-east winds in the southern hemisphere. These winds are known as trade winds.

(c) *The subtropical anticyclones.* Ridges of high pressure between about latitudes 30°–40° N and S, associated with light, variable winds.

(d) *The westerlies.* Belts of generally westerly winds, south-west in the northern hemisphere and north-west in the southern hemisphere, between about latitudes 40° and 60°.

(e) *The temperate-latitude low-pressure systems.* Variable winds converging into low-pressure belts at about 60° N and 60° S.

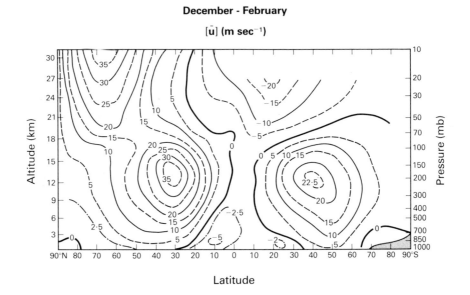

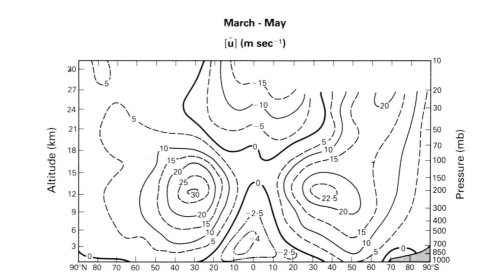

Figure 4.17   Mean zonal winds for the four seasons.

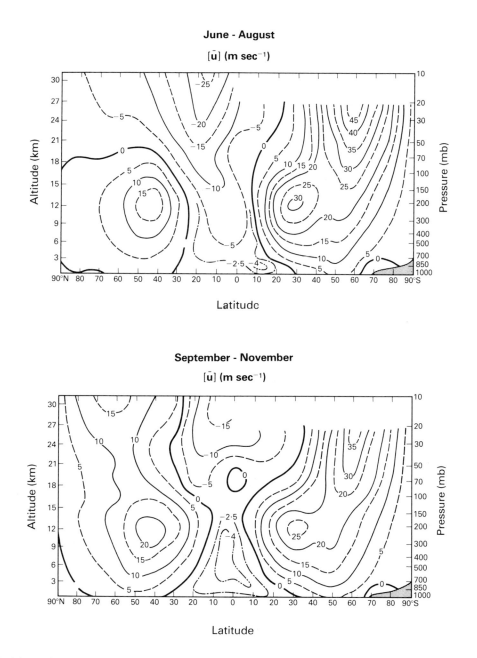

June - August
[ū] (m sec⁻¹)

September - November
[ū] (m sec⁻¹)

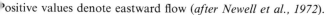

Positive values denote eastward flow (*after Newell et al., 1972*).

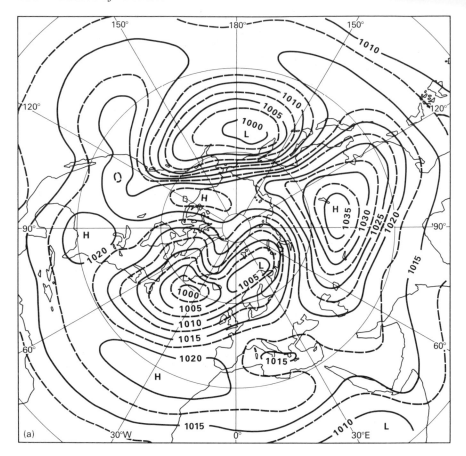

Figure 4.18   (a) Average pressure for January 1951–66;

(f) *The polar anticyclones.* Regions of outflowing winds with an easterly component, diverging from weak high-pressure systems near the poles.

Recent investigations by Newell *et al.* (1972) have shown that the classical picture of the meridional circulation shown in Figure 4.16 is oversimplified. Figure 4.19, which is based on research carried out by Newell *et al.*, shows estimates of the mean meridional circulation (north–south) for the four seasons. The meridional patterns for the mid-seasons clearly correspond to the classical picture, but during the extreme seasons the Hadley cell of the winter hemisphere dominates the circulation pattern of the tropics and the summer Hadley cells of the northern hemisphere almost disappears. The mean circulation for the extreme seasons is largely the result of the monsoon circulation over Asia and the Indian Ocean. During the northern summer the normal tropical circulation pattern is reversed over Asia, with rising air in the subtropics and a high-level southward flow that destroys the sub-

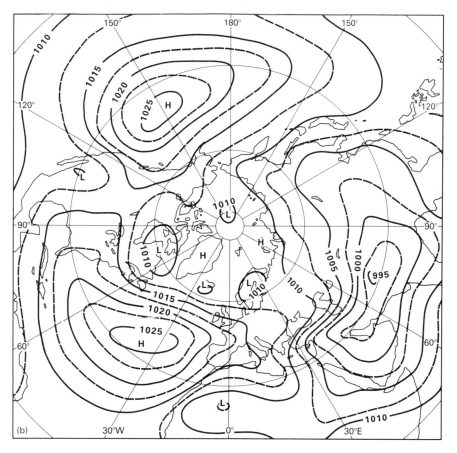

(b) average pressure for July 1951–66 (*after Lamb et al., 1973*).

tropical westerly jet stream and creates an easterly jet stream near the equator. Two weak Hadley cells exist over the Pacific and Atlantic Oceans during this season but the mean meridional circulation is completely dominated by the flow over Asia. Similarly, in winter the mean pattern is determined by the circulation over Asia. Thus a Hadley cell exists in each hemisphere, even in the extreme seasons, at all longitudes except those occupied by the Asiatic monsoon, and the results shown in Figure 4.19 arise from averaging a rather weak Hadley cell motion extending over four-fifths of the tropics with a strong monsoonal flow in the remaining fifth.

3.4 The thermal Rossby number

The laboratory experiments which most nearly duplicate the basic motions observed in the atmosphere are those involving a rotating annulus of fluid subjected to axisymmetric heating and cooling. The actual apparatus can

December - February

Mass flux 10¹² g sec⁻¹

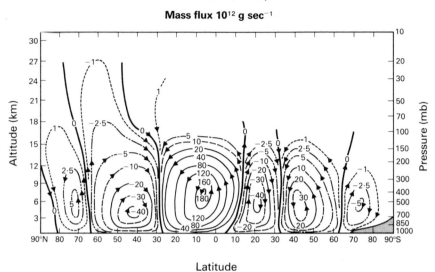

Latitude

March - May

Mass flux 10¹² g sec⁻¹

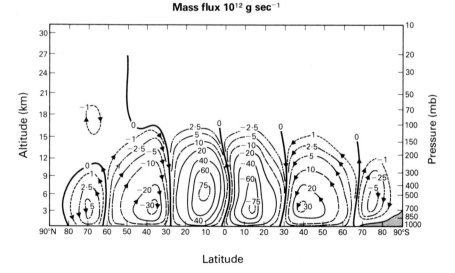

Latitude

Figure 4.19   Mean meridional circulations for the four seasons (*after Newell et al., 1972*).

June - August

Mass flux 10¹² g sec⁻¹

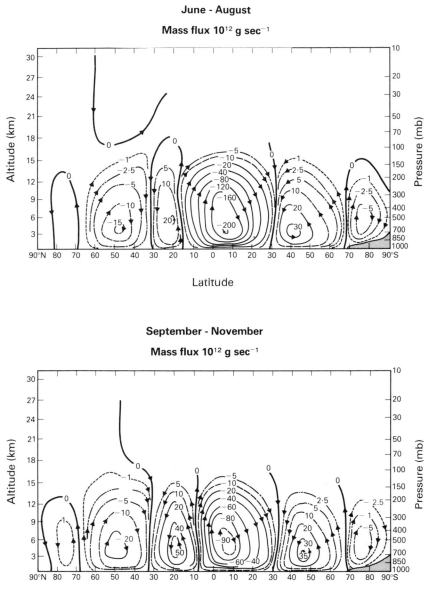

Latitude

September - November

Mass flux 10¹² g sec⁻¹

Latitude

assume a variety of forms, but basically it consists of a convection chamber formed by the annular space between two concentric cylinders of which the vertical walls can be either heated or cooled. The whole apparatus is rotated on a turntable, and the resulting motions in the fluid, which can be a mixture of water and glycerol, are made visible by the use of tiny neutrally-buoyant reflecting solid particles such as polystyrene beads. The particle motions are recorded by a camera which rotates with the convection chamber, and so a time exposure will record particle motions relative to the rotating apparatus as streaks. Typically the radius of the outer surface of the inner cylinder varies from 0–4 cm, while the inner surface of the outer cylinder may have a radius of about 9 cm and the fluid depth is about 16 cm, while rates of rotation vary from 0 to about 7 rad s^{-1}.

Though various experimental arrangements are possible, the one of greatest interest in the context of the general circulation is that in which the central cylinder acts as a cold source (pole) while the outer cyclinder is a warm source (equator). If the temperature gradient is kept constant and the rates of rotation varied, a variety of interesting flow patterns become apparent. At low rates of rotation the flow is completely symmetrical about the axis of rotation, but as the rotation rate increases, the flow becomes non-axisymmetric and waves appear which are arranged around the central axis. The fluid flows through the waves in the direction of rotation, with the flow concentrated into narrow bands or streaks, since the velocity is not uniform over the whole wave. It is also observed that the waves progress around the axis in the direction of rotation, and that the flow exhibits regular periodic time-variations, including a change in the number of waves, and shows the regular progression around the wave pattern of sizeable distortions and a wavering in the shape of the flow pattern. As the rate of rotation increases so does the number of waves until eventually the flow becomes irregular.

Flow patterns observed in the annular container appear to be controlled by both the thermal contrast and the rate of rotation, that is to say, by the so called thermal Rossby number. The thermal Rossby number (R) depends upon the temperature contrast (ΔT) in the annulus and the inverse of the rate of rotation (Ω), resulting in the following:

$$R \propto \frac{\Delta T}{\Omega^2}$$

From the annulus experiments it appears that non-axisymmetric flow occurs with low thermal Rossby numbers, that is with relatively high rates of rotation, while axisymmetric flow occurs with high thermal Rossby numbers. The vertical component ($\Omega \sin \varphi$) of the earth's rotation about its axis is dependent upon latitude, being zero at the equator and a maximum at the pole. Thus the middle latitudes could be considered as corresponding to the experiments with small thermal Rossby numbers (large rates of rotation) and the equatorial latitudes to those with large thermal Rossby numbers (small rates of rotation). Different types of atmospheric circulations are observed in tropical and temperate latitudes, the latter exhibiting waves and jet streams in the upper atmosphere which are very similar to those observed in the annulus experiments. Thus part of the difference between the equatorial and middle-latitude atmospheres is due to the variation of the Coriolis parameter with latitude.

3.5 The subtropical anticyclones

Two belts of high pressure (Figure 4.18) at about 30° N and 30° S contain several quasi-permanent anticyclones known as the subtropical anti-cyclones, separated from each other by cols. The most notable ones in the northern hemisphere are the two oceanic highs, in the Pacific and the Atlantic respectively, and the North African high, which fails to show up at sea-level but emerges clearly at the 3 km level. In the southern hemisphere semi-permanent anticyclones are found over the Pacific and Indian Oceans.

The subtropical anticyclones which tend to be located at fixed positions on the globe, show greatest permanence and only slight seasonal variations. With regard to the geographical equator, the subtropical high-pressure ridges reflect a slight asymmetry, the southern ridge being situated 5° latitude closer to it in the mean than the northern one. They show only small seasonal displacements, amounting to about 5° latitude on average, as they are nearest to the equator in winter. At the subtropical ridge lines, mean pressure is practically equal in both hemispheres, varying on average from 1,015 mbar in summer to 1,020 mbar in winter.

Subtropical anticyclones are formed by air subsiding from high levels in the atmosphere, in this case out of the subtropical jet streams and, typically, the air may take up to three weeks to subside from 12 km to 3 km. Because the air is sinking it is warming adiabatically, and therefore the anticyclones are cloud-free in the middle layers. The continual subsidence in the subtropical anticyclones makes it almost impossible for extensive clouds to form and therefore for rain to fall. The important arid zones of the world are found around latitudes 30° N and 30° S, where large areas are dominated by anticyclones all the year round. Situated in these latitudes are most of the large deserts of the world including the Sahara in North Africa, the Arabian and Syrian deserts, Death Valley in North America, the Atacama Desert in South America, the Kalahari Desert in southern Africa and the Australian deserts. The great subtropical rainless zones spread over the oceans as well as the land masses for the lack of rainfall is not due to the absence of surface water.

A good approximation (Smagorinsky 1963) to the subtropical anticyclone latitude at the surface as well as the latitude of the middle-latitude/Hadley transition aloft is given by φ in the equation:

$$\tan \varphi = -\frac{H}{R}\left(\frac{\partial\theta/\partial z}{\partial\theta/\partial y}\right)$$

where H is a scale height, roughly the height of the 500 mbar level of the atmosphere; R is the radius of the earth; $\partial\theta/\partial z$ is the vertical lapse rate of potential temperature; and $\partial\theta/\partial y$ is the north–south gradient of potential temperature.

Logarithmic differentiation of this equation shows that a 1 per cent change in either the vertical lapse rate or the horizontal temperature gradient changes the tangent of φ by 1 per cent. For summer subtropical anticyclones, this is about 0·2° of latitude. An increase in the horizontal temperature gradient leads to a decrease in the latitude of the subtropical anticyclones and this is shown in Figure 4.20.

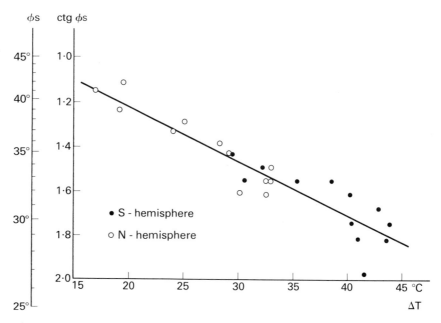

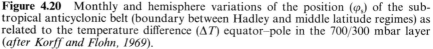

Figure 4.20   Monthly and hemisphere variations of the position (φ_s) of the sub-tropical anticyclonic belt (boundary between Hadley and middle latitude regimes) as related to the temperature difference (ΔT) equator–pole in the 700/300 mbar layer (*after Korff and Flohn, 1969*).

Miles and Follard (1974) have plotted non-overlapping means for 5-year periods from 1900–74 for the North Atlantic region (10° W to 60° W) of the latitudes of the subpolar minimum of pressure, the subtropical maximum of pressure, and the latitude of strongest westerly winds. The results are shown for the whole year and for the summer (June, July and August) in Figure 4.21. The annual diagram shows no trend or even a slight one towards the pole in the latitude of the subtropical pressure maximum. The summer diagram shows a clear drift of the subtropical pressure maximum towards the pole. Miles and Folland comment that results for the whole hemisphere are similar.

3.5.1   *The deserts of the Sahara, Near and Middle East*
An arid belt extends from the Atlantic off the western coast of Africa right into the heart of the Asiatic continent, i.e. over a zonal section of about 110° longitude. According to Flohn (1964), the causes of the aridity of this

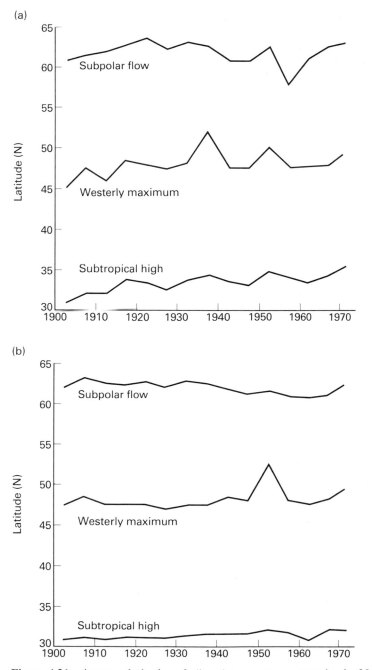

Figure 4.21 Average latitudes of climatic zone parameters in the North Atlantic. Five-year means except the last plot which is 1970–73. (a) Summer; (b) whole year (*after Miles and Folland, 1974*).

vast area during summer are by no means clear. This becomes obvious when the extent of the summer rains in this arid belt is compared with similar rains in other continents. Figure 4.22 illustrates that in the Sahara and Arabia, tropical summer rains extend only, as a regular phenomenon, to about 16–18° N, and in the form of irregular events up to about 22° N. In contrast, in southern Africa the tropical summer rains cover southern Rhodesia and the Kalahari semi-desert; they extend right into the centre of South Africa and merge with the prevailing summer rains in the belt of westerlies.

In Australia the tropical summer rains extend at least to 24–26° S, and occasionally to 30° S. The large zonal dimension of Australia (40° longitude) is smaller than that of northern Africa (68° longitude), but this is hardly sufficient to explain the striking difference in the latitudes reached by the

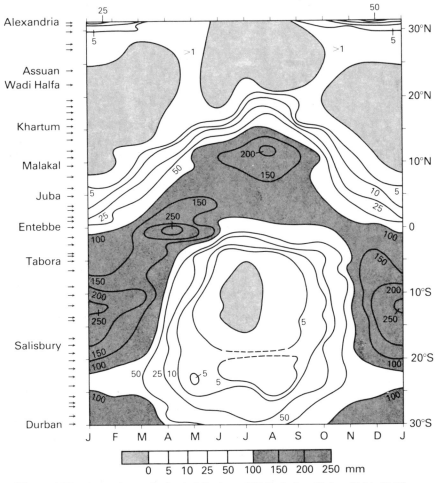

Figure 4.22 Annual trend of rainfall along 32° E. (*after Flohn, 1964, 1966*).

rains. As a consequence the average summer rainfall remains at least 100 mm, even in the centre of the great Australian desert, while in the broad central belt of the Sahara (including Arabia) the average amount of rainfall remains below 20 mm.

Similarly, in Mexico and the south-western United States a merger of tropical summer rains with the continental summer rains is observed extending right into the centres of the Arizona heat low. On the continent of South America, the merger of the tropical summer rains with the rains within the westerlies is well established.

According to Flohn (1964) the unique summer aridity of the desert belt from the western Sahara to Pakistan, is strongly correlated with the forced descending motion on the northern side of the tropical easterly jet stream (Figure 4.23). During the northern summer, the tropical easterly jet stream extends in the layer 200–100 mbar in the latitude belt 5–20° N from the Philippines across southern Asia and northern Africa to the western Atlantic, i.e. over nearly half of the earth's circumference. The high wind speeds above Aden and Khartoum certainly support the use of the term 'jet stream'; at both stations values above 50 m s^{-1} are not infrequent. Above central and western Africa the gradual decrease of wind speed is clearly visible in the streamlines shown in Figure 4.23. In this whole exit area the very gradual deceleration of the jet stream core results in widespread convergence aloft and sinking motions on the northern side. Over Africa a meridional cell is observed with sinking motions on the subtropical northern flank and rising motions on the equatorial side. Thus the deceleration of the easterly jet stream intensifies the aridity of the deserts of North Africa and the Middle East. The reverse flow is observed in the entrance region over south-east Asia, where air sinks to the south and rises to the north, with a rainy area over southern Asia.

Fairbridge (1976) considers that the present Sahara desert is a relatively youthful feature. The paleo-equator of the late Cretaceous passed diagonally across North Africa, through the vicinity of Cairo. During the Cenozoic there was a progressive plate motion, in a counter-clockwise sense, bringing Africa steadily northwards. Thus the Tropic of Cancer, which had lain over western Europe in the late Cretaceous came over North Africa in the late Tertiary. Marine Cretaceous transgressions across the Sahara joined the Niger delta to the Mediterranean and there were widespread Tertiary continental basins where fluvio-lacustrine deposits accumulated. According to Fairbridge it is only in the late Quaternary that there is evidence of eolian facies.

Figure 4.24 suggests that the Würm/Wisconsin glaciation was marked by hyperaridity in the Sahara. During this period, Saharan sands reached as far south as the Zaire basin and from the southern hemisphere Kalahari sands migrated northward to cross the Zaire River near Kinshasa. These old dunes are today largely buried beneath tropical vegetation and are partly degraded. Fairbridge considers that the dune sands of the Sahara evolved progressively during the Pliocene and Quaternary, expanding over greater and greater areas with each glacial desiccation, but interrupted to a large

Streamlines 200 m

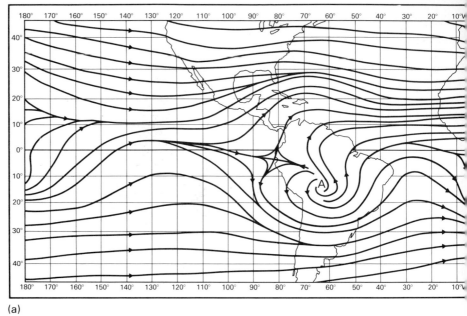

(a)

Streamlines 200 m

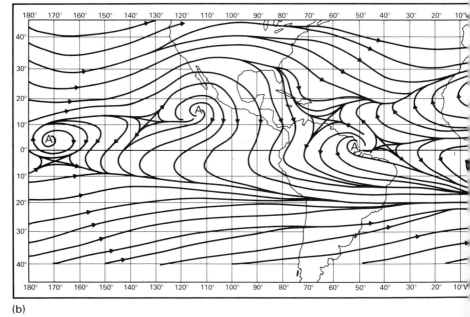

(b)

Figure 4.23   200 mbar streamlines for: (a) December–February; and (b) June–Aug

cember - February

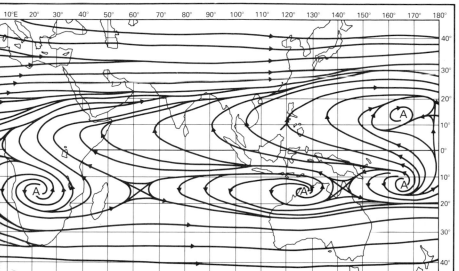

ne - August

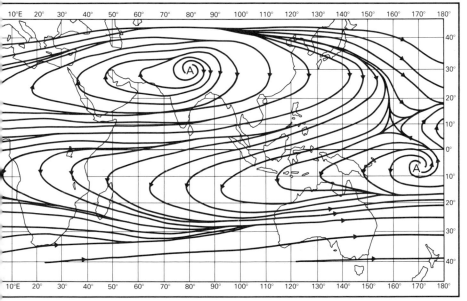

he subtropical and equatorial easterly jet streams are clearly visible (*after Newell et al., 1972*).

extent during each interglacial phase when there were widespread humid conditions that involved valley development and lake growth.

Figure 4.24 shows that following the glacial aridity there was a late-Pleistocene phase of very heavy precipitation in central and subtropical Africa, which began around 13,000 BP and continued until about 8,000 BP. Since this pluvial period there has been a general trend towards increased aridity. Similar long-term trends are observed in the other semi-arid regions of the world. In particular it has been observed in the Kalahari region, eastern Brazil, western India, Iran, Central Asia and Australia.

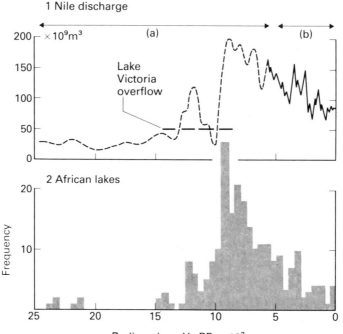

Figure 4.24   Hydrology in tropical Africa.
(1) Nile discharge as recorded by sedimentary accumulation of silt and by 'Nilometer' measurements by the Pharoahs.
(2) Histogram of 238 radiocarbon dated records of high lake levels from intertropical Africa. The lakes include those of Tibesti, Chad, the Niger oases, Mauritanian sebkhas, Hoggar, East Africa (rift valley), Ethiopia and Sudan, but exclude Lake Victoria and others related to the Nile; thus there is no direct connection between diagrams (1) and (2) except by common meteorological history (*after Fairbridge, 1976*).

3.6 The trade winds

Trade winds occur on the equatorial side of the subtropical anticyclones and occupy the bulk of the tropics. Broadly speaking, the trade winds blow from ENE in the northern hemisphere and from ESE in the southern. There

Plate 6   Rock paintings from Tassili-n-Ajjer, Algeria, showing a great number of cattle. The paintings date from between 2000 and 4000 BC and suggest that the Sahara was wetter at this particular time. *Photo by Henri Lhote, reproduced by permission of Hutchinsons.*

tends to be a certain monotony about the weather of the trade winds since the steadiness of the trades reflects the permanence of the subtropical anti-cyclones, which are inclined to be most intense in winter, making the trade winds strongest in winter and weakest in summer. All tropical oceans have extensive areas of trade winds except the northern Indian Ocean.

During the summer of 1856 an expedition under the direction of C. Piazzi-Smyth visited the island of Tenerife in the Canary Islands to make astro-nomical observations from the top of the Peak of Tenerife. On two of the journeys up and down the 3,000 m mountain, Piazzi-Smyth carefully measured the temperature, moisture content, wind direction and speed of the local trade winds. He found that an inversion was often present and that it was not located at the top of the north-east trade regime, but was situated in the middle of the current; thus it could not be explained as a boundary between two air-streams from different directions. He also noticed that the top of the cloud layer corresponded to the base of the inversion. These observations by Piazzi-Smyth have been confirmed many times and the trade-wind inversion is now known to be of great importance in the meteorology of the tropics.

Broad-scale subsidence in the subtropical anticyclones is the main cause of the very dry air above the trade-wind inversion. The subsiding air normally meets a surface stream of relatively cool maritime air flowing towards the equator. The inversion forms at the meeting point of these two air-streams, both of which flow in the same direction, and the height of the inversion base is a measure of the depth to which the upper current has been able to penetrate downward.

Subsidence is most marked at the eastern ends of the subtropical anti-cyclonic cells, that is to say along the desert cold-water coasts of the western edges of America and Africa, for it is here that the trade inversion is at its lowest. Normally as the trade winds approach the equator, the trade inversion increases in altitude and conditions become less arid. The typical cloud of the trade winds is the cumulus cloud, which is formed by warm bubbles or 'thermals' rising from the surface. The sky between the cumulus clouds is clear, and so large amounts of solar radiation reach the surface. Over the oceans the intense radiation evaporates water which is carried aloft by the thermals and eventually distributed throughout the layer below the trade inversion. The result is that the layer below the inversion becomes more moist as the trade wind nears the equator and the continual convection in the cool layer forces the trade inversion to rise in height.

3.7 The equatorial trough

Trade winds from the northern and southern hemispheres meet in the equa-torial trough, which is a shallow trough of low pressure, generally situated on or near the equator. Over the oceans it lies in the belt of the doldrums and has a north and south movement which follows the sun with a timelag of one or two months. Latitudinal changes normally amount to only a few degrees during the year, except over the Indian Ocean where they amount to about 30°. Though the winds are calm or light easterly in many equatorial regions, semi-permanent equatorial westerlies are found over the eastern Indian Ocean, Indonesia and Malaysia.

Cloud and rain in the equatorial trough are often associated with the inter-tropical convergence zone (ITCZ) which is a relatively narrow zone into which air-streams converge. Over the sea it occurs at or very near the latitude of maximum surface temperature, which is usually some distance off the equator, while over land it forms at the boundary between converging air-streams. Inspection of individual photographs from earth satellites show that the ITCZ can only rarely be identified as a long, unbroken band of heavy cloudiness. Rather, it is usually made of a number of 'cloud clusters', separated by large expanses of clear sky. The cloud clusters are marked by heavy rainfall (>2 cm day^{-1}) and normally move slowly westward. Thus, in a climatological sense, the ITCZ may be considered to be the locus of westward propagating cloud clusters. Both single and double versions of the ITCZ are known, sometimes one convergence zone is observed near the

Plate 7 Near vertical view of thundercloud over South America from Apollo 9. *Reproduced by permission of the National Aeronautics and Space Administration.*

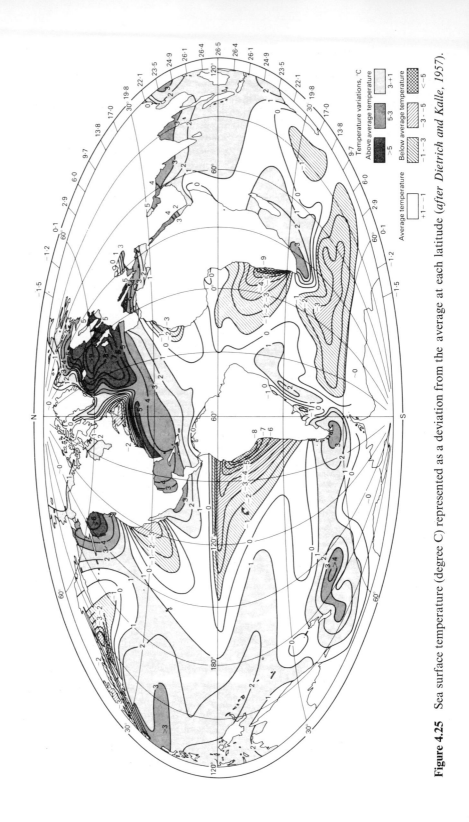

Figure 4.25 Sea surface temperature (degree C) represented as a deviation from the average at each latitude (*after Dietrich and Kalle, 1957*).

equator while at others there are two marking the northern and southern limits of the equatorial trough.

The distribution of mean annual rainfall within the equatorial trough is extremely variable. Rainfall maxima are observed over the equatorial regions of South America and Africa, and also over the equatorial islands from Indonesia to the Carolines. In contrast rainfall minima tend to exist over the oceans. Variations within the trough become clearer if the annual percentage of days with thunder are considered, since thunderstorm peaks are found over South America, Africa and Indonesia, while they are comparatively rare over the oceans. The bulk of the rainfall over the equatorial continents and Indonesia/Malaysia falls from thunderstorms, but this is not true of the oceans.

Dietrich and Kalle (1957) have mapped the difference between the sea surface temperature and its average global value along the part of each latitude circle situated over the oceans. The results are illustrated in Figure 4.25, where cold-water regions are defined as ocean areas with sea surface temperatures below the global average. By far the most extensive is the south Pacific cold water, for it stretches westwards from the coast of South America by about 85° longitude; in contrast the south Atlantic cold water continues westward from the coast of Africa by only about 40° longitude. The temperature difference along most of the coast of Peru exceeds -8 degree C and the cold water reaches almost to the equator where off Equador, differences of -3.5 degree C are found. The corresponding coastal cold water off south-west Africa has an equally large negative temperature difference, but it does not extend very far to the north. The subtropical cold-water areas in the north Atlantic and north Pacific are relatively small, and the north Indian Ocean has only a minor area off Somalia; there is also a minor patch off Australia in the south Indian Ocean.

Bjerknes (1969) has studied the relationship between rainfall and air–sea temperature differences at Canton Island, which lies on the edge of the south Pacific cold water. Upwelling cold water reaches Canton Island for most of the year, but occasionally the upwelling process ceases and the sea surface becomes warmer than the atmosphere. Bjerknes found that large monthly totals of rain at Canton Island occur only during periods when the ocean is warmer than the atmosphere. When the ocean is colder than the atmosphere the rainfall is usually slight. The Canton Island type of rainfall regime—that is, normal aridity but occasional wet seasons—is known to prevail throughout most of the south Pacific cold-water region. Along the coast of Peru the surface moist layer is less than 800 m deep and normally only occasional drizzle falls from stratus cloud. However, during the southern summer the south-easterly trades are sometimes replaced by northerly winds which in turn induce a southward flow of warm equatorial water, displacing the cool upwelling water of the Peruvian coastal current. The warm water and unstable air result in heavy rainfall in an otherwise almost completely arid desert—a phenomenon known locally as the El Niño effect.

Bjerknes considers that when the cold ocean water along the equator is well developed, the air above will be too cold to take part in the ascending

Zonal mass flux (10¹² g sec⁻¹) 5°N, December - February

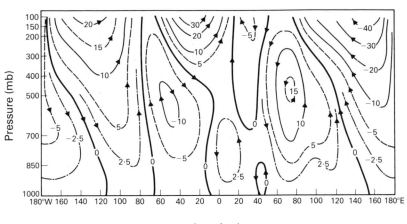

Longitude

Zonal mass flux (10¹² g sec⁻¹) 5°S, December - February

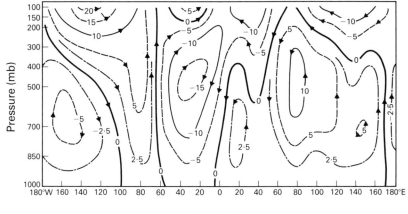

Longitude

Figure 4.26 Zonal air flow near the equator during the mid-season months (*after Newell et al., 1974*).

motion of the Hadley cell circulation. Instead, the equatorial air flows westward between the Hadley cell circulations of the two hemispheres to the warm west Pacific, where, having been heated and supplied with moisture from the warmer waters, the equatorial air can take part in large-scale, moist adiabatic ascent. In Indonesia huge cloud clusters with a diameter of more than 600 m develop each day, giving an area-averaged rainfall amount around 2,200 mm per year, equivalent to a release of latent heat of about

Zonal mass flux (10¹² g sec⁻¹) 5°N, June - August

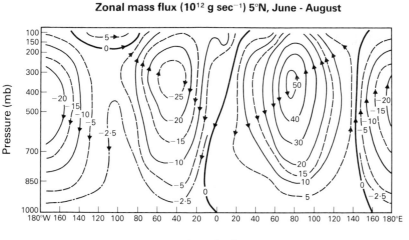

Longitude

Zonal mass flux (10¹² g sec⁻¹) 5°S, June - August

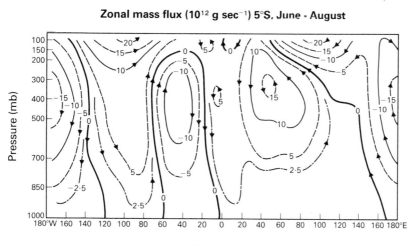

Longitude

170 W m⁻², which is much more than the net radiation near the surface. This thermally-driven circulation (Figure 4.26) between an equatorial heat centre—the 'maritime continent' of Indonesia—and a cooling area in the eastern Pacific is often known as a Walker circulation. As shown in Figure 4.25, the equatorial Atlantic is analogous to the equatorial Pacific in that the warmest part is in the west, at the coast of Brazil, but west–east contrasts of water temperature are much smaller than in the Pacific. However, in

January (Figure 4.26) a thermally-driven Walker circulation may operate from the Gulf of Guinea to the Andes, with the axes of the circulation near the mouth of the Amazon.

The strength of the Walker circulation varies with the surface water temperature of the Pacific ocean and in turn forms part of a wider tropical fluctuation known as the 'southern oscillation'. According to Berlage (1966), 'the southern oscillation is a fluctuation of the intensity of the intertropical general atmospheric and hydrospheric circulation. This fluctuation, dominated by an exchange of air between the south-east Pacific subtropical high and the Indonesian equatorial low, is generated spontaneously. The period varies between roughly 1 and 5 years and amounts to 30 months on the average.' In general terms, when pressure is high in the Pacific Ocean, it tends to be low in the Indian Ocean from Africa to Australia; these conditions are associated with low temperatures in both these areas, and rainfall varies in the opposite direction to pressure (Walker and Bliss 1932).

Troup (1965) and Bjerknes (1969) have interpreted the surface pressure, sea surface and air temperatures, cloudiness, and mean geopotential height data over the Pacific as implying a Walker circulation in that region that is an integral, if not controlling, part of the southern oscillation. Generally, the equatorial Pacific easterlies result in upwelling of cold water along the South American coast and in the equatorial region stretching toward Canton Island. There is an east–west gradient of sea surface temperature, with the cold water in the eastern and central Pacific contrasting with the warm sea surface temperatures in the western equatorial Pacific. In association with this, the surface pressure is relatively high over the cold eastern Pacific and relatively low over the warm western Pacific. This pressure gradient tends to maintain strong equatorial easterlies which in turn maintain the cold upwelling so that the process is a self-sustaining or even an accelerating one.

In years when the equatorial upwelling process is interrupted, the equatorial east–west sea surface temperature gradient is much reduced, the pressure gradient is weakened so that the equatorial easterly flow is weakened, the unusually warm sea surface temperatures feed latent and sensible heat into the atmosphere, which promotes the Hadley circulation; the north-east trades are at a maximum, and rainfall and cloudiness are more prevalent in the eastern Pacific.

According to Bjerknes, the differences in the two aspects of the circulation are most pronounced in the December–February season. This is illustrated in Table 4.1, where circulation characteristics for December to February in two contrasting years are listed.

3.8 Middle and high latitudes

An indirect circulation cell, that is to say sinking warm and rising cool air, is observed in middle latitudes. Temperature falls poleward in middle latitudes, but not in a uniform manner, because horizontal temperature gradients are concentrated in narrow 'frontal zones'. These frontal zones are also the main regions of ascent in the middle-latitude cell, and air ascending poleward over fronts should attain large eastward components if its absolute angular

Table 4.1 Circulation characteristics for two successive years (*after Newell et al., 1974*).

Parameter	December 1962–February 1963	December 1963–February 1964
Sea surface temperatures at Canton Island	cold	warm
Surface pressure anomaly		
Canton Island	+1·2 mbar	−0·8 mbar
Djakarta	−0·6 mbar	+0·3 mbar
Rainfall anomaly		
Canton Island	−166 mm	+469 mm
Djakarta	+569 mm	−409 mm
Vertical motion −500 mbar		
Eastern Pacific	weak	strong
Indian Ocean	strong	weak
Hadley circulation	weak	strong
Walker circulation	strong	weak
Eddy momentum fluxes	weak	strong

momentum is conserved. Hence in middle latitudes the general flow is towards the east and jet streams in the upper troposphere are closely associated with frontal zones near the surface. The average meridional circulation in middle latitudes does not appear to be capable of transporting all the heat and momentum required by balance considerations, and so instead, part of this transport is carried out by horizontal eddies. These eddies take the form of large meanders in the upper troposphere westerlies and closed circulations at the surface.

Compared to the Hadley cell the middle-latitude atmosphere is highly disturbed and the suggested meridional circulation in Figure 4.16 is largely schematic. The average positions of the very large-scale meanders or waves in the upper troposphere appear to be geographically fixed but the minor waves are more mobile. Near the surface, closed circulations exist in the form of depressions and anticyclones which drift slowly towards the east. The depressions normally contain the frontal zones along which much of the middle-latitude ascent takes place, so the feature marked as the polar front in Figure 4.16 is highly variable in character and position. The brief picture sketched above for the middle latitudes compares well with that conveyed by rotating fluid annulus experiments with low thermal Rossby numbers. It appears that the disturbed weather of the middle latitudes arises partly from the high values of the Coriolis parameter found in these regions, and not completely from the large thermal contrasts.

Both polar regions are located in regions of general atmospheric subsidence, though the climate is not particularly anticyclonic and the winds are not necessarily easterly. The moisture content of the air is low because of the intense cold, and horizontal thermal gradients are normally weak, with the result that energy sources do not exist for major atmospheric disturbances, which are rarely observed. Vowinckel and Orvig (1970) suggest that the arctic atmosphere can be defined as the hemispheric cap of fairly low kinetic energy circulation lying north of the main course of the planetary westerlies, which places it roughly north of 70° N.

Long-term mean circulation charts for the middle latitudes show only a faint resemblance to individual synoptic charts. The long-term mean charts are composed of averages of circulation patterns for a great variety of scales and types and do not reveal the traits of individual synoptic situations. A study of daily charts has disclosed that middle-latitude circulation patterns undergo a series of irregular quasi-cyclic changes over a period of weeks, which are best analysed in terms of the zonal index and the index cycle. The strength of the zonal circulation is conveniently measured by the mean pressure difference between two latitude circles, which is called the zonal index, for sea-level conditions 35° and 55° N are often used.

A typical index cycle can be divided into four stages (Figure 4.27) with the following characteristics:

1　an initial high zonal index with strong sea-level westerlies and a long-wavelength pattern in the upper atmosphere;

2　an initial lowering of the sea-level zonal index with an associated shortening of the wavelength pattern aloft;

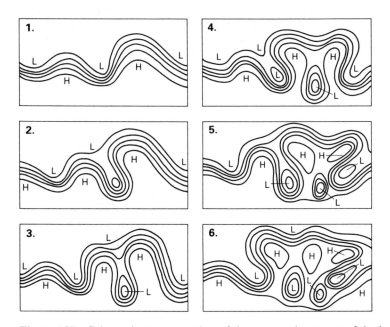

Figure 4.27　Schematic representation of the consecutive stages of the breakdown of a zonal flow in the 500 mbar surface; the development takes place during a period of eight to ten days and the chart can be assumed to extend from the Mississippi Valley in the west to the Ural Mountains in the east (*after Rossby, 1959*).

3   the lowest sea-level zonal index characterized by a complete break-up of the sea-level zonal westerlies with closed cellular centres, and with a corresponding breakdown of the wave pattern aloft;

4   finally, an initial increase of the sea-level zonal index with a gradual increase of the westerlies and the development of an open wave pattern aloft.

Since the high index situation is characterized by strong latitudinal temperature gradients in middle latitudes, and by little air-mass exchange, the cyclones and anticyclones of the mid-latitude belt drift eastward, often with considerable speed. Under conditions of low zonal index, a warm cut-off high forms in the 1,000–500 mbar thickness pattern, and sea-level cyclones have a tendency to be steered around these highs, either to the north of the cut-off high or to the south along the base of the pattern. Because the normal westerly current is block and reversed, this last arrangement is often called a 'blocking pattern' or simply a 'block'. Anticyclonic development is associated with the warm ridge and sometimes cyclonic development with the cold trough of the blocking pattern, and so an extensive blocking anti-cyclone often exists in middle and high latitudes, while a non-frontal low exists to the south. Blocking patterns once formed are relatively stable and usually persist for a period of 12–16 days.

Recent trends in the climate of the middle and high latitudes are shown in Figures 4.21 and 4.28. Figure 4.21 shows that both the latitudes of the north Atlantic westerlies and the north Atlantic subpolar flow drifted slightly northwards in the early 1970's, after having moved south in the late 1950's and early 1960's. The 1,000–500 mbar thickness (Figure 4.28) reveals a more or less steady decrease for the period since 1959 for the area north of 50° N,

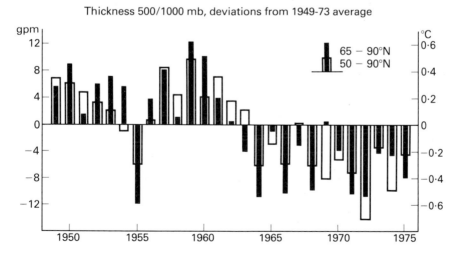

Figure 4.28   Variation of the average annual temperature of the 1,000/500 mbar layer of the polar cap. The thickness of the 1,000/500 mbar layer, representing the lower half mass of the atmosphere, is directly proportional to the vertically average temperature, including a small humidity correction (*after Flohn, 1977*).

but the Arctic has warmed in the mid-1970's, especially in the region from the Norwegian Sea across the Russian Arctic to near the Bering Strait. Nevertheless, according to Painting (1977) the annual mean temperatures deduced from mean 1,000–500 mbar thicknesses show a net cooling over the hemisphere north of 40° N since about 1970 which amounts to an average value near to 0·2 degree C per decade.

4 The hydrological cycle

The full cycle of events through which water passes in the earth–atmosphere system is best illustrated in terms of the hydrological cycle (Figure 4.29), which describes the circulation of water from the oceans, through the atmosphere back to the oceans, or to the land and thence to the oceans again by overland and subterranean routes. Water in the oceans evaporates under the influence of solar radiation and the resulting clouds of water vapour are transported by the atmospheric circulation to the land areas, where precipitation may occur in the form of rain, hail or snow. Some of this precipitation will infiltrate into the saturated zone beneath the water-table, from where it flows slowly through aquifers to river channels or sometimes directly to the sea. The water remaining on the surface will partly evaporate back to the atmosphere and partly form surface run-off into rivers, which eventually run into the oceans. To the moisture which has evaporated directly back to the atmosphere from the soil surface, must be added a further loss of water in the soil via transpiration through plants. Not all the above stages will necessarily occur in any particular example of the hydrological cycle or at any particular place or time. During droughts the cycle may appear to have stopped, whereas during floods it may seem to be continuous, and yet both of these phenomena can occur in the same place at different seasons.

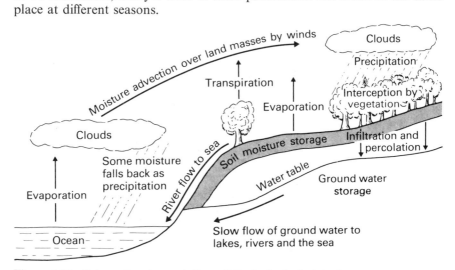

Figure 4.29   Schematic representation of the hydrological cycle.

It is simplest to regard the hydrosphere as a global system consisting of four reservoirs interrelated by the flux terms of the hydrological cycle (SMIC report 1971). The four reservoirs are the world ocean, polar ice, terrestrial waters, and atmospheric waters, and estimates of the amount of water stored at the present time are summarized in Table 4.2. The sum of the four storage components is constant since there is negligible creation or destruction of H_2O. For comparison, the water balances of the continents and oceans are shown in Tables 4.3 and 4.4. Residence times of water in the various stores are shown in Table 4.5.

Table 4.2 Review of estimations on principal components of the hydrosphere (total water resources of the globe $= 1 \cdot 5 \times 10^6$ km^3) (*after SMIC Report, 1971*).

	Area covered (10^6 km^2)	Present amount of water resources		Changes of the amount during the last	
		(10^3 km^3)	(per cent)	18,000 years (10^3 km^3)	80 years (10^3 km^3)
World ocean	360	1,370,000	93	+40,000 (100 m in level)	+100 (0·27 m in level)
Polar ice	16	24,000	2		+40
Terrestrial waters	134	64,000†	5		
Atmospheric waters	510	13	0·001		

† Includes groundwaters 64,000; lakes 230; soil moisture 82.

Table 4.3 Average water balance of the land areas (cm yr^{-1}) (*after Budyko, 1971*).

	Precipitation	Evaporation	Run-off
Africa	69	43	26
Asia	60	31	29
Australia	47	42	5
Europe	64	39	25
North America	66	32	34
South America	163	70	93
Total land areas	73	42	31

Table 4.4 Average water balance of the oceans (cm yr^{-1}) (*after Budyko, 1971*).

	Precipitation	Run-off	Evaporation	Flow
Atlantic	89	23	124	−12
Indian	117	8	132	−7
Pacific	133	7	132	+8
World ocean	114	12	126	0

Table 4.5   Average residence times of water (*based on SMIC Report, 1971*).

Atmospheric waters	10 days
Terrestrial waters	
Rivers	2 weeks
Lakes	10 years
Soil moisture	2–50 weeks
Biological waters	few weeks
Groundwaters	up to 10,000 years
Polar ice	15,000 years
World ocean	3,600 years

On average, the water vapour in the air at any one time represents about one-fortieth part of the annual rainfall or about ten days' supply of water. At any given time the earth's atmosphere contains on average an amount of water vapour, which if it were all condensed and deposited on the surface of the earth, would stand to a depth of about 25 mm. The amount of water held in the atmosphere, at any given time, in the form of water vapour is small compared with the storage capacities of the soil, deep rocks and oceans, and so the main hydrological function of the atmosphere is to transport water vapour; in fact, over a number of years there is an exact global balance between evaporation and precipitation.

The amounts of water held in storage in the various reservoirs are not constant and have varied during geological time. In particular, the ice ages caused major redistributions of water storage. Water stored in ice-sheets cannot form run-off to the oceans and so causes a corresponding decrease in ocean storage. The formation of the Antarctic ice-sheet in the late Tertiary caused a worldwide lowering of sea-level by about 55 m. The low sea-level of 18,000 BP (see Figure 5.11) was associated with the increase in land ice masses during the Würm/Wisconsin glaciation. According to Orvig (1972), mean sea-levels for the period 1890 to 1940 rose by about 0·2 m, but since 1940 the rate of rise appears to have decreased by about 40 per cent. According to Orvig, recent studies of the Antarctic ice budget indicate that it is slightly positive at present, with an annual gain of perhaps a little less than 600 km^3 of water. Greenland has a small net annual loss of a little less than 100 km^3 of water. The net result should be a lowering of sea-level of approximately 1·5 mm yr^{-1}. This simultaneous occurrence of a positive mass budget in Antarctica together with a rise in world sea-level requires an explanation. Flohn (1972) suggests that it represents an increase in temperature and therefore of volume of the world ocean. The SMIC report (1971) suggests that some of the water may have come from non-recharged groundwater resources.

4.1 Vertical motion and the formation of precipitation
Precipitation which is the source of all fresh water on the earth's surface, may take the form of rain, snow, hail, frost or dew. Over most of the earth's surface the most important form of precipitation is rainfall, though dew

may be important locally in arid regions and snowfall on high mountains and in polar regions. Individual storms contribute widely differing amounts of precipitation to the earth's surface, but the percentage of annual precipitation supplied by large and small storms appears to be fairly constant irrespective of location on the globe, for it is found that approximately half of the annual precipitation over any given area is contributed by about one-quarter of the storms experienced. In contrast, about half the storms experienced contribute about one-quarter of the annual precipitation total. Most of the annual precipitation total is caused, therefore, by a comparatively small number of major storms, while the great majority of storms produce rainfalls which are not hydrologically significant.

Four mechanisms are necessary for the production of significant precipitation:

(a) a lifting mechanism to produce cooling of the air;

(b) a mechanism to produce condensation of water vapour and formation of cloud droplets;

(c) a mechanism to produce growth of cloud droplets to sizes capable of falling to the ground against the rising air currents implied by condition (a);

(d) a mechanism to produce sufficient accumulation of water vapour in the storm to account for the observed precipitation rates.

4.1.1 *Cooling and vertical motion*

Vertical ascent of air accompanied by adiabatic cooling is the only known mechanism capable of producing the degree and rate of cooling needed to account for heavy rainfall. Thus the variation of precipitation in space and time is largely determined by the spatial and temporal variation in the vertical motion of the air. The vertical motions observed within the atmosphere are largely a result of the dynamic processes within the atmosphere itself and the interactions of the atmosphere with the underlying surface. They are associated with weather systems of various scales ranging from thunderstorms to tropical cyclones and temperate-latitude depressions. Vertical air motions in the atmosphere which give rise to significant rainfall may be classified as convective, banded and general.

Convective vertical motion occurs in cells with a diameter of about 1·5 to 8 km and a vertical extent from about 1,000 to more than 15,000 m. These cells are visible in the atmosphere as cumulus and cumulonimbus clouds and give rise to the shower type of precipitation. It is a characteristic of convective activity that the resulting precipitation is local in extent and of only short duration. Over relatively featureless terrain the spatial development and movement of convective cells is essentially random, but when the surface has marked orographic relief there is a tendency for convective cells to develop over preferred areas.

Banded vertical motion, which occurs over a zone from a few kilometres to about 80 km in width and several hundred kilometres in length, is usually associated with fronts and convergence zones, or is the result of ascent over a mountain by an air-stream.

General vertical motion is associated with large-scale weather systems such as tropical cyclones. Broad-scale ascent takes place in these systems and the resulting precipitation can extend over the area of a medium-sized country. Organized convection bands commonly occur within tropical cyclones but the isohyetal patterns for the complete duration of such storms frequently lack significant cellular features. Storms of this type can have life cycles of many days and travel many thousands of kilometres.

4.1.2 *Formation of precipitation*

The cooling associated with rising air parcels produces a relative humidity of 100 per cent, and further cooling will produce supersaturation. It has been found in the laboratory that pure, dust-free air can have a relative humidity of considerably above 100 per cent without condensation, while it is observed that condensation takes place in the atmosphere with very small supersaturations, implying that atmospheric air differs in some way from pure, dust-free laboratory air. Investigations have shown that condensation in the atmosphere takes place on small dust particles having an affinity for water, known as condensation nuclei. The origin of these condensation nuclei is mostly dust from the soil surface, salt particles formed by the evaporation of sea spray and industrial smoke. There are always sufficient nuclei present in the lower atmosphere for condensation to occur if the air is cooled to saturation, and in smoky industrial areas before the saturation point is reached.

It is further found that while impure water freezes in the laboratory at 0 °C, water droplets exist in the atmosphere at temperatures considerably below 0 °C. Liquid water at a temperature below 0 °C is said to exist in the supercooled state, and nearly all clouds which extend above the 0 °C isotherm contain supercooled droplets at some stage in their history. In the atmosphere supercooled water droplets exist at temperatures down to −40 °C, so it is only at very high levels in the troposphere that pure ice-crystal clouds, such as cirrus, are found. Clouds at temperatures between 0 °C and −40 °C are observed to contain mostly supercooled water droplets with just the occasional ice-crystal.

Clouds are essentially suspensions of minute liquid water drops and ice-crystals. The water drops are extremely small being of the order of 1–100 μm in diameter, while the typical raindrop has a diameter of 1 mm. The basic problem is therefore to explain the growth of cloud droplets to such a size that they can fall against the rising air currents and form rain. There are two main theories of this process, one known as the ice-crystal process and the other as the coalescence process.

The ice-crystal process was first suggested by Bergeron and is sometimes called the Bergeron process. He maintained that if a sufficient number of ice-crystals in close proximity to water droplets are present in cloud, that there will be a movement of water vapour from the droplets to the ice-crystals since the saturation vapour pressure over ice is less than that over water. This difference in vapour pressure is largest in the −10 to −20 °C range. The ice-crystals then grow as snowflakes and as they fall they pick up more water by the accretion of small cloud droplets. If the temperature at ground level is at or below 0 °C, the resulting precipitation is in the form of snow, but if the snowflakes fall through warmer air in the lower atmosphere they melt and the resulting precipitation is in the form of rain. This process is a major mechanism for the production of precipitation in the middle latitudes and accounts for both rain and snow, and the observed change from rainfall to snowfall with increasing altitude in mountainous regions.

In the tropics heavy rain is often observed to fall from clouds which do not reach the 0 °C isotherm, which may be as high as 4 or 5 km. Under these conditions the rain cannot possible be caused by the ice-crystal process because the cloud contains no ice, being completely above 0 °C in temperature. Such observations have led to the development of the coalescence theory, which is not competitive but complementary to the Bergeron idea. If the cloud contains some water droplets which are appreciably larger than the great majority of droplets, the slower rate of rise of such large drops in a cloud updraught leads to collisions and, in some cases, coalescence with the smaller droplets. Factors which promote this process are appreciable cloud depth and updraught speed which permit growth by accretion to a size sufficient to ensure that the drop will not evaporate at the top of the cloud but will fall back through the cloud, growing further by collision with small droplets and eventually reaching the ground as rain. The original large droplets probably form on extremely hygroscopic salt particles which in turn originate by the evaporation of spray from the sea surface. Evidence exists that the coalescence process is predominant at temperatures higher than −12 °C, while the ice-crystal process is more important at lower temperatures.

Since condensation nuclei are required for the formation of precipitation, it is a very effective means of removing aerosols from the troposphere. The stratosphere is cloud-free, therefore aerosols have a very long residence time in the stratosphere as compared with the troposphere.

4.1.3 *Inflow of moist air*

Precipitating clouds are organized into weather systems on various scales, and a characteristic feature of these weather systems is marked convergence of air in the lower atmosphere. The inflowing air provides a continual supply of moisture to the weather system and thus allows for the observed rainfalls. For instance, warm moist equatorial air can hold in the form of water vapour the equivalent of about 100 mm of rainfall, but an average equatorial shower will produce rainfall at the rate of 100 mm per hour for several hours, thus implying the need for a continual flow of water vapour into the system. When it is considered that even the most efficient storm can convert only a small amount of the available atmospheric water vapour into precipitation, the problem of moisture flow into the precipitating system is seen to be of some importance. It is observed in many storms that the amount of rainfall produced is limited by the inflow of moist air, since if the inflowing air is dry the rainfall is small, while if it is very moist the rainfall is extremely heavy. Certainly for the production of continuous intense precipitation a strong inflow of very moist air into the storm is an essential condition in all parts of the world.

4.1.4 *Global distribution of precipitation*

On average the water vapour in the air at any one time represents about one-fortieth part of the annual rainfall, or about ten days' supply of water. As rainfall consists of moisture which has been advected into the area by the winds, it is therefore reasonable to expect a marked correlation to exist between wind systems and rainfall. It is observed that the average water vapour content of the atmosphere does not correlate very clearly with annual rainfall, and this is because rainfall is caused by weather systems and storms in the atmosphere. Consequently annual rainfall distributions reflect both the incidence of weather systems and also the atmospheric moisture content.

References

BERLAGE, H. P. 1966: The southern oscillation and world weather. *Mededeelingen en Verhandelingen* **88**. The Hague: Korunklijk Nederlands Meteorologisch Instituut.

BJERKNES, J. 1969: Atmospheric teleconnections from the equatorial Pacific. *Monthly Weather Review* **97**, 163–72.

BUDYKO, M. I. 1974: *Climate and life*. New York: Academic Press.

DIETRICH, G. and KALLE, K. 1957: *Allgemeine Meereskunde*. Berlin: Gebrüder Borntraeger.

FAIRBRIDGE, R. W. 1976: Effects of holocene climatic change on some tropical geomorphic processes. *Quaternary Research* **6**, 521–56.

FLOHN, H. 1964: Investigations on the tropical easterly jet. *Bonner Meteorologische Abhandlugen* **4**. Bonn.

1966: Warum ist die Sahara trocken? *Zeitschrift für Meteorologie* **17**, 316–20.

1972: The hydrological cycle of Greenland and Antarctica. In IASH, *World water balance* **3**, 665. Gentbrugge.

1977: Climate and energy: a scenario to a 21st century problem. *Climatic Change* **1**, 5–20.

HANSON, K. J. 1976: A new estimate of solar irradiance at the earth's surface on zonal scales. *Journal Geophysical Research* **81**, 4435–43.

KORFF, H. C. and FLOHN, H. 1969: Zusammenhang zwischen dem Temperatur-Gefälle Äquator – Pol und den planetarischen Luftdruckgurteln. *Annalen der Meteorologie Neue Folge* **4**, 163–4.

LAMB, H. H., COLLISON, P. and RATCLIFFE, R. A. S. 1973: Northern hemisphere monthly and annual mean sea-level pressure distribution for 1951–66, and changes of pressure and temperature compared with those of 1900–39. *Geophysical Memoirs* **118**. London: Meteorological Office.

MILES, M. K. and FOLLAND, C. K. 1974: Changes in the latitude of the climatic zones of the northern hemisphere. *Nature* **252**, 616.

MOFFITT, B. J. and RATCLIFFE, R. A. S. 1972: Northern hemisphere monthly mean 500 millibar and 1000–500 millibar thickness charts and some derived statistics (1951–66). *Geophysical Memoirs* **117**. London: Meteorological Office.

NEWELL, R. E., KIDSON, J. W., VINCENT, D. G. and BOER, G. J. 1972: *The general circulation of the tropical atmosphere* **1**. Cambridge, Mass.: MIT Press. 1974: *The general circulation of the tropical atmosphere* **2**. Cambridge, Mass.: MIT Press.

ORVIG, S. 1972: The hydrological cycle of Greenland and Antarctica. In IASH, *World water balance* **3**, 664–5. Gentbrugge.

PAINTING, P. J. 1977: *A study of some aspects of the climate of the northern hemisphere in recent years*. Scientific Paper **35**. London: Meteorological Office.

PALMEN, E. 1951: The role of atmospheric disturbances in the general circulation. *Quarterly Journal Royal Meteorological Society* **77**, 337–54.

RASCHKE, K., VONDER HAAR, T. H., BANDEEN, W. R. and PASTERNAK, M. 1973: The annual radiation balance of the earth–atmosphere system during 1969–70 from Nimbus 3 measurements. *Journal Atmospheric Sciences* **30**, 341–64.

ROSSBY, C. G. 1959: Current problems in meteorology. In BOLIN, B., editor, *The atmosphere and sea in motion*, 9–50. New York: Rockefeller Institute Press.

ROTTY, R. M. and MITCHELL, J. M. JR 1974: *Man's energy and world climate*. Institute for Energy analysis report, Oak Ridge Associated Universities.

SCHNEIDER, S. H. with MESIROW, L. E. 1976: *The genesis strategy: climate and global survival*. New York: Plenum Press.

SMAGORINSKY, J. 1963: General circulation experiments with the primitive equations (Appendix B). *Monthly Weather Review* **91**, 159–62.

SMIC REPORT 1971: *Inadvertent climate modification. Report of the study of man's impact on climate*. Cambridge, Mass.: MIT Press.

TROUP, A. J. 1965: The southern oscillation. *Quarterly Journal Royal Meteorological Society* **91**, 490–506.

VOWINCKEL, E. and ORVIG, S. 1970: The climate of the north polar basin. In ORVIG, S., editor, *Climates of the polar regions*. Amsterdam: Elsevier.

WALKER, G. T. and BLISS, E. W. 1932: World Weather V. *Memoir Royal Meteorological Society* **4**, no. 36. London.

5
The Evolution of Surface Climates: Glacial and Interglacial Periods

1 The feasibility of climatic change

A few years ago, Lorenz (1968) raised a fundamental question: Is our climate stable? Non-linear systems, which are far simpler than the atmosphere, sometimes display a tendency to fluctuate in an irregular manner between two or more internal states, while the external boundary conditions remain completely unchanged. This behaviour is related to the system's transitivity and is illustrated in Figure 5.1. Assume that two different states of a climatic system are possible at a time $t = 0$, such as A and B in Figure 5.1, and let A be the climatic state that would normally be 'expected' under the given constant boundary condition. In a completely transitive system, the climatic state B would approach the state A with the passage of time and eventually become indistinguishable from it. This would correspond to a unique solution

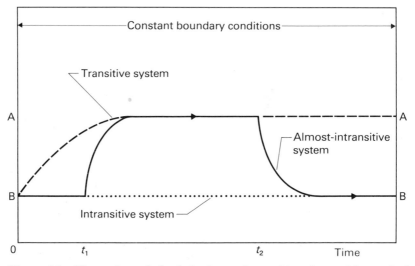

Figure 5.1 Illustration of the behaviour of transitive, intransitive and almost-intransitive climatic systems with respect to an initial climatic state B. The climatic state A is an alternative bistable state under the same boundary conditions (*after US National Academy of Sciences, 1975*).

for the climate under fixed boundary conditions. In a completely intransitive system, on the other hand, the climatic state B would remain unchanged, and two possible solutions would exist. There would be no way in which it would be possible to identify the state A as the 'normal' or correct solution, as state B would presumably furnish an equally acceptable set of climatic statistics.

A third type of behaviour is however possible, and is displayed by an almost-intransitive system. In this case, the system in state B may behave for a while as though it were intransitive, and then at time t, shift toward an alternative climatic state A, where it might remain for a further period of time. At time t_2 the system might then return to the original climatic state B, where it will remain or enter into further excursions. The climate exhibited by such a system would thus consist of two quasi-stable states, together with periods of transition between them.

At least two types of climate are possible under existing conditions. The first is the present observed world climate while the second is the so called 'white earth' solution. Under white earth conditions the whole of the global surface is covered by ice and snow. While the calculated temperature depends considerably on the choice of albedo for the ice-covered earth's surface, the mean latitudinal temperatures remain below zero for any plausible albedo. Suggested temperature values range from -68 °C at the equator to -73 °C at the pole. Thus the 'white earth' is a second type of climate which is possible under existing conditions. There is also a possibility of the existence of a third variant of world climate, involving partial glaciation of the earth and lower temperatures as compared to the actual ones.

On the assumption that climatic changes are more than just random fluctuations, paleoclimatologists have long sought evidence of regularities in proxy records of the earth's climatic history. For example, many aspects of the global ice fluctuations during the last 700,000 years may be summarized in terms of a 100,000 year cycle, as shown in Figure 5.2. Each such period is marked by a gradual transition from a relatively ice-free climate (or interglacial) to a short, intense glacial maximum and followed by an abrupt return to an ice-free climate. So on the time-scale of some thousands of years the earth's climate acts as an almost-intransitive system.

Evidence of an almost-intransitive climate during the 300-year period of instrumental records is very difficult to find. Evidence for this only really exists on a much longer time-scale. Moving averages of climatic data do show non-periodic oscillations, but only in a few cases is some kind of a flip–flop mechanism at work, producing a bimodal or multimodal distribution. Of special interest in this connection is the extreme variance of rainfall in the equatorial regions of the Pacific and its positive correlation with surface water temperature.

Changes in rainfall patterns occur nearly simultaneously along the 12,000 km equatorial sector from the Ecuador coast (80° W) to Nauru (169° E) and are closely related to the El Niño phenomenon along the coasts of Peru–Ecuador. The physical cause of the rainfall changes is found in the change of sign of the helical Ekman drift of the ocean layer above the

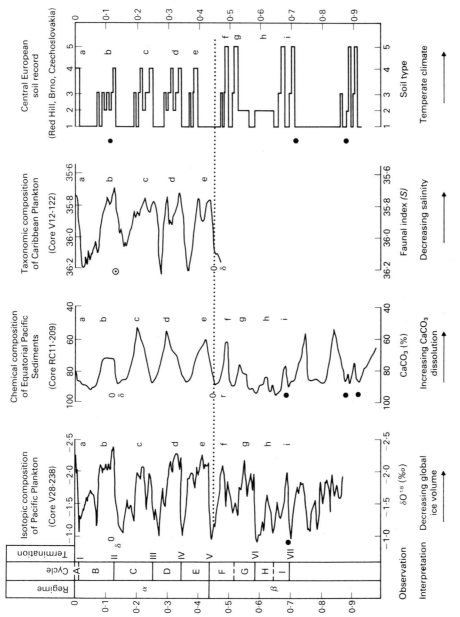

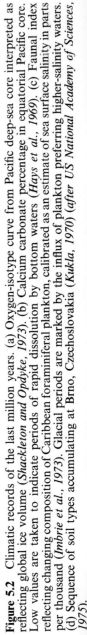

Figure 5.2 Climatic records of the last million years. (a) Oxygen-isotype curve from Pacific deep-sea core interpreted as reflecting global ice volume (*Shackleton and Opdyke, 1973*). (b) Calcium carbonate percentage in equatorial Pacific core. Low values are taken to indicate periods of rapid dissolution by bottom waters (*Hays et al., 1969*). (c) Faunal index reflecting changing composition of Caribbean foraminiferal plankton, calibrated as an estimate of sea surface salinity in parts per thousand (*Imbrie et al., 1973*). Glacial periods are marked by the influx of plankton preferring higher-salinity waters. (d) Sequence of soil types accumulating at Brno, Czechoslovakia (*Kukla, 1970*) (*after US National Academy of Sciences, 1975*).

thermocline on both sides of the equator. While normally equatorial up-welling causes stabilization together with cloudless arid conditions, a weakening of the south-east trade system leads to equatorial downwelling, with tropical warm water, destabilization and cumulonimbus convection with humid conditions. The processes are illustrated in Figure 5.3 and described in detail in Chapter 4. Equatorial upwelling and aridity can be considered as

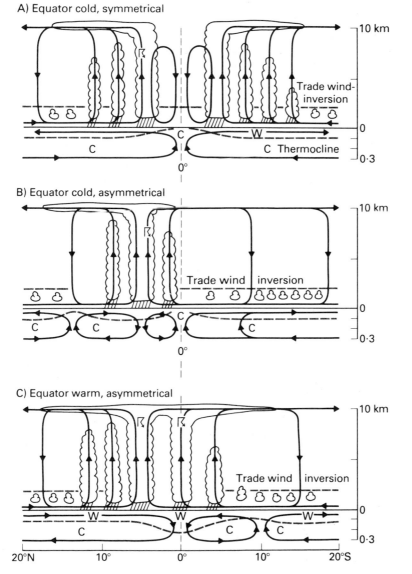

Figure 5.3 Meridional circulation patterns of troposphere and upper ocean mixing layer in the equatorial Pacific (*after Flohn, 1975*).

normal, while during the irregular anomaly periods the oceanic evaporation increases by at least 20–30 per cent and the rainfall by a factor of 5 to 20, producing large-scale effects on the atmospheric circulation. Thus Flohn (1975) has commented on a short-lived but drastic El Niño episode in the equatorial Pacific, which occurred simultaneously with the peak of the Sahel drought of the early and mid-1970's, with a severe deficiency of monsoon rains in the Indian subcontinent and with an exceptional drought in the European area of the Soviet Union. A preliminary survey by Flohn (1975) suggests a nearly simultaneous occurrence of such anomalies in different parts of the Tropics (Figure 5.4), indicating some kind of time-dependent zonal circulation as described by Bjerknes (1969) as the 'Walker circulation'.

A question of some importance is the shape of the continuous variance spectrum of climatic fluctuations. A uniform distribution of variance as a function of frequency (or 'white noise') would imply a lack of predictability in the statistical sense or a lack of 'memory' of prior climatic states. A 'red noise' spectrum, on the other hand, in which the variance decreases with increasing frequency, implies some predictability in the sense that successive climatic states are correlated. An initial estimate of the variance spectrum of temperature has been made from the fluctuations on time-scales from 1 to

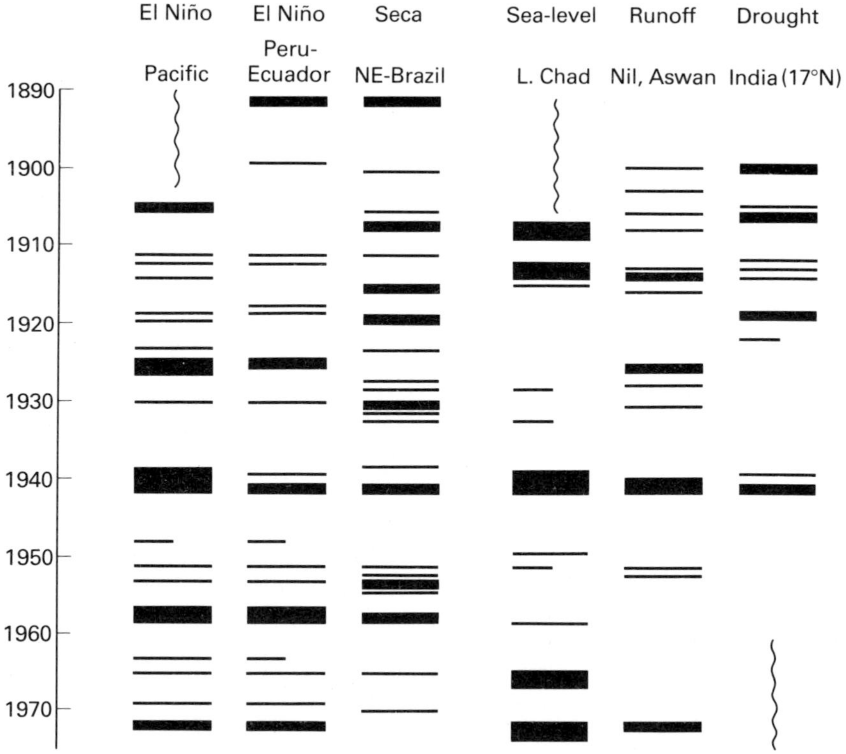

Figure 5.4   Synopsis of climatic anomalies in equatorial regions (*after Flohn, 1975*).

10,000 years by Kutzback and Bryson (1974). The most significant feature of the spectrum is that it is very red at low frequencies (periods greater than 1,000 years), red at intermediate frequencies and nearly white at high frequencies (periods less than 10 years). The existence of non-zero inter-correlations in such a spectrum implies that some portion of the climatic system retains a 'memory' of prior states. In view of the relatively short memory of the atmosphere, it seems likely that this is provided by the oceans on time-scales of years to centuries and by the world's major ice-sheets on longer time-scales.

2 Chronology of global climate

The long-term paleoclimatic history of the earth is not well known, but it appears that in the very remote past the earth evolved into a temperature regime differing little from that of the present day. So it is believed that the pattern of climate with tropical regions, cool poles and periodic ice ages has altered scarcely at all at least during the last 10^9 years. As geological exploration continues so records of glaciations of considerable severity are announced from ancient rocks covering many distant parts of the earth. However, it is interesting to note that during the last 500 million-year period the earth was warmer than it is today and during more than 90 per cent of the time the poles were ice-free. During the Mesozoic era (200–60 m.y. BP) the annual mean surface temperatures were 8–10 °C near both poles and 25–30 °C in the tropics. Nevertheless, some evidence is available which suggests several extensive glaciations, the oldest and most widespread one beginning before the Cambrian Era, 550 m.y. BP. Another glacial period occurred during the Permo-Carboniferous (about 300 m.y. BP), at a time when the earth's land masses were joined in a single supercontinent known as Pangaea. The area of this continent was distributed in roughly equal proportions between the hemispheres, with a concentration of land in mid-latitudes. Glaciated portions of Pangaea included parts of what are now South America, Africa, India, Australia, and Antarctica, which at the time occupied southern high-latitude positions. This particular glacial period probably lasted for 30 to 50 m.y.

It is clear that the Permo-Carboniferous ice-sheet covered many thousands of square kilometres at different stages during its evolution. A study of ice-flow directions, as plotted on a reconstruction of Gondwanaland, shows that the evolution of the Permo-Carboniferous glaciation apparently followed that inferred for the present-day glaciation of Antarctica; that is to say there were the initial formation of local ice-caps in highland regions and these then extended laterally to coalesce and form a true ice-sheet. As with the Pleistocene ice-sheet in North America and in Europe, this ice-sheet undoubtedly advanced and retreated following fluctuations in climatic conditions, but there remains no firm evidence for such variations.

During the last part of the Mesozoic era (from 100 m.y. to 65 m.y. BP) global climate was in general substantially warmer than it is today, and the polar regions were without ice-caps. From the climatic viewpoint, the most striking aspect of world geography 100 m.y. BP was the essentially meridional configuration of the continents and shallow ocean ridges, which must have prevented a circumpolar ocean current in either hemisphere. According to Dietz and Holden (1970) this barrier (Figure 5.5) was formed by South America, Antarctica, Australia, and by the narrow and relatively shallow ancestral Indian Ocean. About 50 m.y. years ago the Antarctica–Australian passage began to open, and as India and Australia moved northward, the Indian Ocean widened and deepened. According to paleontological and sedimentary evidence the Antarctic circumpolar current system was first established about 30 m.y. BP. The US National Academy of Sciences (1975) consider that this was one of the pivotal events in the climatic history of the last 100 m.y.

There is some evidence of a world temperature maximum during the Cretaceous and early Tertiary, with mean global temperatures higher than at any time since the Carboniferous glaciations. About 55 m.y. BP numerous geological records make it clear that global climate began a long cooling trend sometimes known as the Cenozoic climatic decline. Kennett (1977) has reconstructed the history of the Antarctic continent during this period of climatic decline. Paleomagnetic evidence shows that the Antarctic continent has essentially been in a polar position since at least the late Cretaceous. Initial glaciation followed by ice-cap formation, however, did not commence until much later in the middle Cenozoic, demonstrating that a near-polar position is not sufficient for glacial development. During the early Eocene, Australia began to drift northward from Antarctica, forming an ocean, although the circum-Antarctic flow was blocked by the continental South Tasman Rise and Tasmania. During the Eocene the Southern Ocean was relatively warm and the continent largely non-glaciated, with cool temperate vegetation existing in some regions. The first major climatic–glacial threshold was crossed about 38 m.y. ago near the Eocene–Oligocene boundary, when substantial Antarctic sea-ice began to form, resulting in a rapid temperature drop in bottom waters of about 5 degree C. The deep oceanic circulation initiated at this time was much like that of the present day. This climatic threshold was crossed as a result of the gradual isolation of Antarctica from Australia and perhaps the opening of the Drake Passage. According to Kennett, widespread glaciation probably occurred throughout Antarctica in Oligocene times (38 to 22 m.y. BP), but no ice-cap existed. By the middle to late Oligocene (30 to 25 m.y. BP), deep-seated circum-Antarctic flow had developed south of the South Tasman Rise, as this had separated sufficiently from Victoria Land, Antarctica. The next principal climatic threshold was crossed during the middle Miocene (14 to 11 m.y. BP) when the East Antarctic ice-cap formed causing world sea-levels to fall by up to 59 m. Since the middle Miocene the East Antarctic ice-cap has remained a semi-permanent to permanent feature exhibiting some changes in volume. The most important of these occurred during the latest Miocene (about 5 m.y. BP), when ice volumes increased beyond those of the present day. This event was related to global climatic cooling, a rapid northward movement of about 300 km of the Antarctic convergence, and a eustatic sea-level fall that may

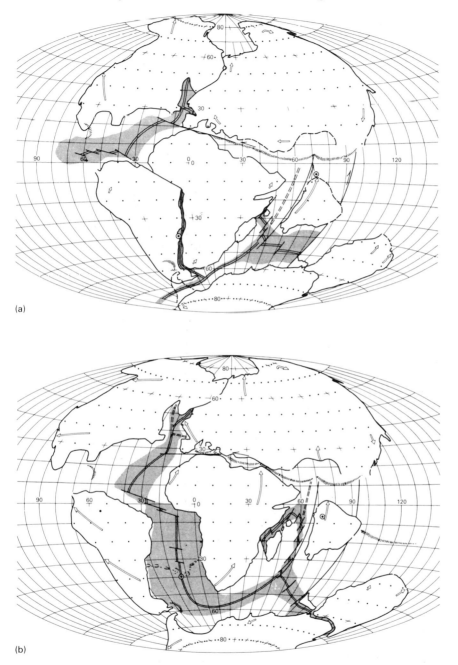

(a)

(b)

Figure 5.5 (a) The continental distribution during the late Jurassic, 135 m.y. BP. (b) The continental distribution at the end of the Cretaceous, 65 m.y. BP (*after Dietz and Holden, 1970*).

have been partly responsible for the isolation of the Mediterranean basin. Although evidence exists for late Miocene cooling in the northern hemisphere, no evidence exists for ice-sheet development until the late Pliocene, about 2·3–3 m.y. BP, when a further global climatic threshold was passed. Since then, major oscillations have continued in the northern hemisphere ice-sheets, forming the classical Quaternary glacial and interglacial episodes.

Figure 5.2 shows that for at least the last 1 m.y. the earth's climate has been characterized by an alternation of glacial and interglacial episodes, marked in the northern hemisphere by the waxing and waning of continental ice-sheets and in both hemispheres by periods of rising and falling temperatures. Continental ice-sheets probably first appeared in the northern hemisphere about 3 m.y. BP, occupying lands adjacent to the North Atlantic Ocean. Figure 5.2 shows that these fluctuations are found in a number of proxy data records. These include the chemical composition of Pacific sediments, fossil plankton in the Caribbean, and the soil types in Central Europe. These 'cycles' identified as A to E by Kukla (1970), may be grouped into a climatic 'regime' covering the last 450,000 years designated α in Figure 5.2. The earlier records (regime β) show higher-frequency fluctuations with less coherence among the various proxy climatic records. Hays *et al.* (1976) have found climatic cycles of 23,000 42,000 and approximately 100,000 years in south hemisphere ocean-floor sediments during the last 450,000 years, suggesting a connection with the Milankovitch mechanism. The dominant cycle is one of about 100,000 years and is seen in the growth and decay of the continental ice-sheets. Indeed, continental ice-sheets appear to develop over a period of 90,000–100,000 years and to terminate rather quickly in about 10,000 years. The interglacial periods between major glaciations lasted about 10,000 years, and were probably never warmer than at the present.

The climatic history of the last 120,000 years is well illustrated by the North Greenland ice-core shown in Figure 5.6. In this diagram temperature is expressed in terms of the amount of oxygen-18 found in the ice. The snow falling over Greenland was formed of water that evaporated from the Atlantic Ocean. Molecules of sea water containing the commoner form of oxygen, oxygen-16, are lighter and evaporate more readily, and remain in vapour form longer than do the small minority that contain heavy oxygen. The proportion of heavy oxygen in the ice therefore gives a general indication of the climate prevailing when the snow fell on Greenland—the less heavy oxygen there is, the colder the climate.

Figure 5.6 shows a warm interval known as the Eemian interglacial followed by a severe glacial period known as the Würm in Europe and the Wisconsin in the United States. The onset of the Würm/Wisconsin glaciation may be dated about 60,000 to 70,000 years ago, when there was a rapid fall in world mean temperatures to near the lowest levels observed in the Würm/Wisconsin glaciation. The cold period following the initial fall in temperature lasted for only a few thousand years and was followed by a relatively warm period lasting till 30,000 years ago. A further cold period followed, lasting about 10,000 years, in which the lowest temperature and the greatest extent of ice-sheets were attained. This particular cold phase

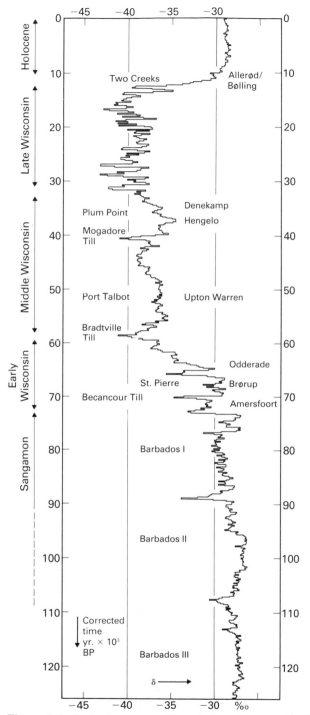

Figure 5.6 Isotopic oxygen-18 variations in a north Greenland ice-core since 130,000 years BP (*after Dansgaard et al., 1971*).

Plate 8   Submerged forest seen at low tide at Ynyslas, Ceredigion, northwest Dyfed, Wales, during spring 1969. There is evidence both of lower sea level and of forest that reached right to the western coasts about 6,000 years ago, when the trees grew. *Photograph reproduced by permission of Dr J. A. Taylor, University College of Wales, Aberystwyth.*

reached its peak between 15,000 and 20,000 years ago. The end of the Würm/Wisconsin glaciation occurred between 13,000 and 10,000 BP, and was marked by a rapid rise of temperature, probably mostly within 2,000 to 3,000 years. Since about 5,000 BP there has been a slight lowering of world mean temperatures.

2.1 Time-scale of climatological changes

A study of the changes in oxygen-18 content in the Greenland ice-core shown in Figure 5.6 suggests that major climatic changes can take place in relatively short time-periods. thus the change from full-glacial to the present inter-glacial climate took place in a period of a few thousand years. Widespread deglaciation began abruptly about 14,000 years ago, and the waning phases of the continental ice-sheets were characterized by substantial marginal fluctuations. The Cordilleran ice-sheet in North America, which had just attained its maximum extent, melted rapidly and was gone by 10,000 BP. The Scandinavian ice-sheet lasted only slightly longer and retreated at the rate of about 1 km per year between about 10,000 and 9,000 years ago. By 8,500 years ago the ice conditions in Europe had reached essentially their present state, and in North America the ice-sheets had shrunk to about their present extent by about 7,000 years ago. The melting of the ice-sheets was not a continuous process, but marked by sudden warmings and coolings. The normal sequence is Bølling interstadial (warm)–older Dryas (cold)–Allerød interstadial (warm)–younger Dryas (cold), covering a period of less than 2,000 years, with variations in annual temperature of up to 6 degree C. The cooling prior to the younger Dryas probably lasted less than 300 yr.

Flohn (1974) has listed a series of sudden coolings in the climatic record. These events show coolings of the order of up to 5 degree C/century in contrast to not more than 1 degree C/century in recent fluctuations. The most dramatic short-lived cooling event was observed in the Greenland ice-sheet record at about 89,000 BP. Here the climate changed from warmer than today into full glacial severity within about 100 yr (Dansgaard *et al.* 1975). This event has also been found in a stalagmite in a French cave at 97,000 BP with a cooling of the cave by 3 degree C in a few centuries and an extremely rapid cooling has been described in many cores from the Gulf of Mexico at 90,000 BP. All the dates are slightly uncertain, hence the differences.

2.2 Historic climatic changes

To obtain an indication of the climate in the northern hemisphere for the last 1,000 years, Lamb (1969) has compiled manuscript references on the character of European weather and has developed an index of winter severity, as shown in Figure 5.7. Lamb's index may be compared with oxygen-18 values preserved in the ice-core taken from Camp Century, Greenland, and mean tree growth at the upper treeline, White Mountains, California. Figure 5.7 suggests a degree of uniformity in the major fluctuations of temperature between the west coast of North America and western Europe.

The early part of the period from about AD 1,100 to 1,400 is sometimes called the Middle Ages warm epoch. This was a period of relatively fine settled climate in Europe, with vineyards existing in southern England. The period from about 1,430 to 1,850 is commonly known as the Little Ice Age, and some records indicate that this period had cold maxima in the fifteenth and seventeenth centuries. Historians differ about the dating of the Little Ice Age, because the long cold period was punctuated by warmer decades and the effects differed from place to place. From the evidence it appears that the atmospheric circulation may have been more meridional than at present, and characterized in western Europe and western North America by short, wet summers and long, severe winters.

Dansgaard *et al.* (1971) has carried out a spectral analysis (Figure 5.8) of the most recent 800 years of the oxygen-18 values from the Camp Century ice-core and found two dominant harmonics with periods of 78 and 181 years. By combining these two harmonics it is possible to produce a smoothed curve which depicts the main trend of the actual observations. The 78-year period has been noticed before in oscillation in the length of the sunspot cycle. The smoothed curve repeats itself at regular intervals and the period 1,350 to 1,500 is a good analogue for the present century. So if past trends are followed in the near future, the last part of the present century should be colder than normal with a slight warming around the year 2,000. If an

Plate 9   Gathering grapes, probably in an English medieval vineyard. *Reproduced by permission of Professor H. H. Lamb, University of East Anglia.*

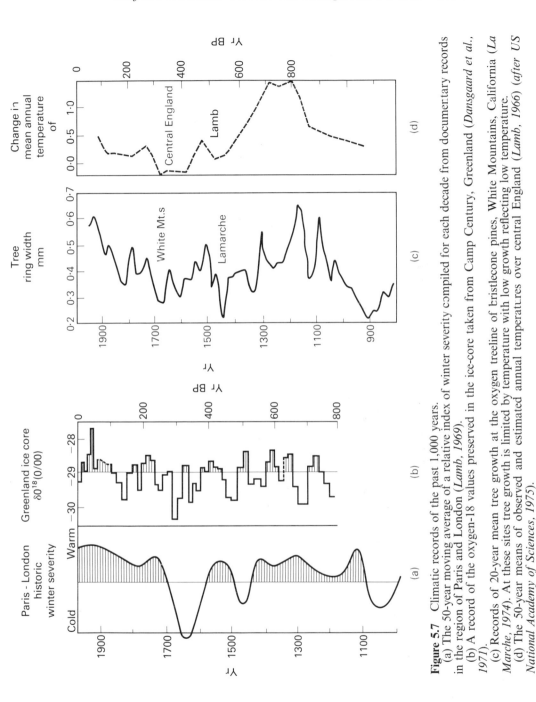

Figure 5.7 Climatic records of the past 1,000 years.
(a) The 50-year moving average of a relative index of winter severity compiled for each decade from documentary records in the region of Paris and London (*Lamb, 1969*).
(b) A record of the oxygen-18 values preserved in the ice-core taken from Camp Century, Greenland (*Dansgaard et al., 1971*).
(c) Records of 20-year mean tree growth at the oxygen treeline of bristlecone pines, White Mountains, California (*La Marche, 1974*). At these sites tree growth is limited by temperature with low growth reflecting low temperature.
(d) The 50-year means of observed and estimated annual temperatures over central England (*Lamb, 1966*) (*after US National Academy of Sciences, 1975*).

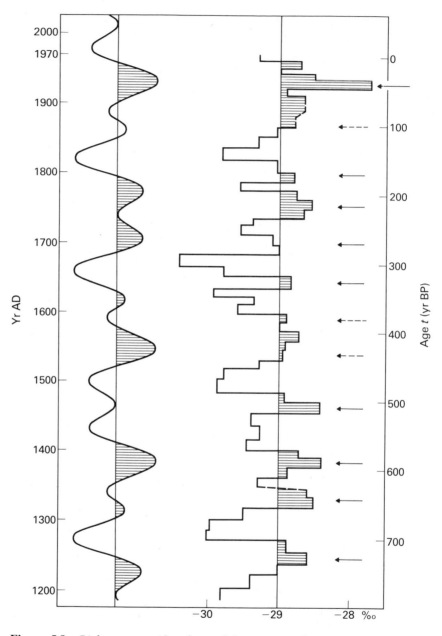

Figure 5.8 *Right:* oxygen-18 values of increments of an ice-core from north Greenland plotted against time *t* since the deposition of the ice. The hatched areas correspond to relatively warm periods.

Left: A synthesis of the two harmonics (78 and 181 years) that dominate the step curve. The curve from 1970 suggests the probable future climatic development (*after Dansgaard et al., 1971*).

Plate 10   Two views of frost fairs on the Thames 1683–4. *Reproduced by permission of the Radio Times Hulton Picture Library.*

analysis is carried out over the last 10,000 years only one other persistent peak appears with a period of 350 years.

Painting (1977) has examined some of the climatic changes over the northern hemisphere in recent years. The 1960's were characterized by excessive blocking in temperature latitudes and a cooling trend over most of the northern hemisphere especially in middle and high latitudes. Some, but not all, of these trends have been reversed in the 1970's. The Arctic has warmed, especially in the region from the Norwegian Sea across the Russian Arctic to near the Bering Strait. This can be seen in both temperature records at high-latitude stations and also in the mean value of the 1,000–500 mbar thickness (see Figure 4.28). Nevertheless, annual mean temperatures deduced from mean 1,000–500 mbar thicknesses show a net cooling over the hemisphere north of 40° N since about 1970 which amounts to an average value near to 0·2 degree C/decade. This has been especially marked south of 60° N where cooling deduced from thickness changes has been confirmed by surface temperature data.

Recent changes in the southern hemisphere have not followed those further north. Damon and Kunen (1976) have recently reported a warming trend at high latitudes in the southern hemisphere.

3 Ice and snow

Terrestrial snow and ice can be divided into five distinct categories which are seasonal snow, permafrost, mountain glaciers, sea-ice, and continental ice-sheets. According to Untersteiner (1975) the total amount of water in all earthly forms is estimated to be $1,384 \times 10^6$ km^3, and of this 97·4 per cent is sea water; 0·0009 per cent is atmospheric vapour, 0·5 per cent is ground-water, most at great depths; 0·1 per cent is contained in rivers and lakes, and 2 per cent is frozen. This last figure means that nearly 80 per cent of the fresh water on earth exists in the form of ice and snow. Today, perennial ice covers 11 per cent of the earth's land surface and an average of 7 per cent of the world ocean. The residence time of solid precipitation in ice masses ranges from 1–10 years for the frozen sea water in the Southern and Arctic Oceans to 10,000–100,000 years in the Antarctic ice-sheet.

3.1 Seasonal snow

The global distribution of snow is shown in Figure 5.9, where it is seen that at the maximum extent in January it covers an area considerably greater than that of all sea-ice and continental ice-sheets combined. There is mounting evidence that at least some of the major glaciations during the Quaternary era began more or less simultaneously in regions where a small anomaly in the area of snow cover might have initiated a positive feedback mechanism, i.e. increased albedo leading to troposphere cooling and a cold upper vortex leading to increased snowfall. Kukla (1975) claims that the northern high- and mid-latitude continents are such sensitive areas, and that the autumn is the sensitive season.

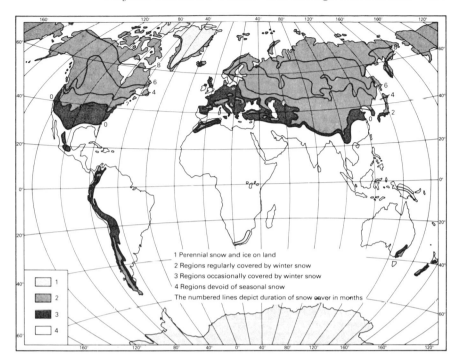

Figure 5.9 Global distribution of snow on land (*after Untersteiner, 1975*).

The results of investigations of snow area for January and February by Painting (1977) are shown in Figure 5.10, where curves (a) and (b) give the total cover, curves (c) and (d) that for 0–180° E and (e) and (f) for 0–180° W. Figure 5.10 shows that there has been no general increase in the maximum snow-covered area in recent years, rather the trend has been for a decrease in the peak area. Similarly, Sanderson (1975) has reported a general decrease in the areas of northern hemisphere sea-ice. Kukla and Kukla (1974) have reported an excessive snow and ice cover for the winter of 1971–2, but there has been little sign since 1972 of large snowfalls over extensive parts of the northern hemisphere. Painting considers that the variability of the total snow-covered area probably amounts to some 15 per cent. He also considers that any attempt to fit trends to existing data are premature and unlikely to provide useful indications of future climatic regimes.

3.2 Sea-ice

At its maximum seasonal extent sea-ice covers nearly 10 per cent of the world ocean, with an average thickness of 3 m in the Arctic and 1 m in the Southern Ocean. In the central Arctic Ocean, a sharp pycnocline exists at a depth of about 25–50 m, forming the lower boundary of the Arctic surface water, with salinities between $32^0/_{00}$ and $34^0/_{00}$ and temperatures close to freezing. The priming factor maintaining this low salinity is the 3,300 km^3 continental

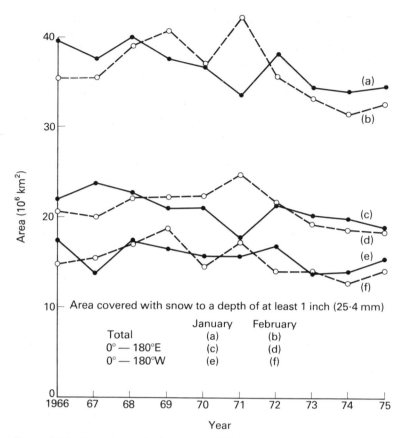

Figure 5.10   Snow cover in January and February 1961–75 over the northern hemisphere (excluding China) (*after Painting, 1977*).

run-off which adds a layer of approximately 30 cm of fresh water to the Arctic Ocean each year. In contrast in the Southern Ocean, with its less persistent ice cover, higher wind velocities, more intense mixing, and absence of continental run-off, the pycnocline is only about one-tenth of the magnitude of that in the Arctic Ocean and is found at variable depths down to 200 m.

Recently, Sanderson (1975) has reported a general decrease in the areas of northern hemisphere sea-ice. Averaging the February, March, April, August and September areas with 7/10 or more sea-ice over two consecutive 3-year periods, namely 1969–71 and 1972–4, gave mean decreases in the latter over the former of 3, 0·3, 3, 8 and 3 per cent for the respective months.

3.3 Glaciers and ice-sheets
The distribution of glaciers is a complex function of the distribution and amount of accumulating snow and of the nature and amount of incoming

energy and its utilization at the glacier surface during the summer ablation period. Thus a particular ice mass exists where, on the average, the amount of snow accumulation equals or exceeds the amount of summer ablation. Temporal changes in the amount and character of snow accumulation or energy input affect this balance and result in glacier growth or shrinkage. Thus the warming trend experienced in the northern hemisphere during the early part of this century led to glacier retreat in most areas; conversely, the downward trend of temperature since about 1940 has caused glaciers to slow this retreat or re-advance. Glaciers are found to exist in a wide range of environments, ranging from the maritime temperate regimes of southern Norway or the North American Cascades on the one hand, to the polar deserts of Canada, northern Greenland and Antarctica on the other.

In general three different major types of glacial climate may be distinguished:

(i)   A widely distributed high-polar ice-cap climate having summers below 0 °C, e.g. north Greenland and Antarctica.

(ii) A continental type of tundra climate with cool summers (warmest month below 10 °C), cold winters (coldest month below −8 °C), wide temperature fluctuations, and small precipitation.

(iii) A maritime type with small oscillations of temperature and cool summers, e.g. Iceland.

True polar climates are today found in two contrasting areas, one of which is largely an elevated plateau and the other an ocean. The Antarctic ice-sheet contains about 90 per cent of the present glacial ice and is more than seven times greater in area than its counterpart in Greenland. The Antarctic continent has an area of about 14,000,000 km^2 of which less than 3 per cent is estimated to be free from a permanent ice-sheet. Elevation makes the environment even less inviting because 55 per cent of the surface is above 2,000 m and about 25 per cent is above 3,000 m. In contrast, the Arctic Ocean is almost completely land-locked except for one main access point to the warmer waters of the Atlantic between Greenland and Norway. The central Arctic basin is covered by a thin (few metres thick) but permanent ice-pack which extends to the continents during the winter. This contrasts with the Antarctic ice-sheet which is up to 2,400 m thick in places.

Precipitation over the Antarctic ice-sheet falls entirely in solid form and ranges from 40 g cm^{-2} yr^{-1} in a narrow band along the coast to 5 g cm^{-2} yr^{-1} and less over nearly half of the continent. Because of observational difficulties in measuring accumulation and ice export in Antarctica, an accurate assessment of mass balance is virtually impossible.

The present consensus is that the Antarctic ice-sheet today has a zero or slightly positive mass balance. Evidence for past variations of the ice-sheet is sparse, but marks of inland ice-levels 60 m above the present surface exist. Indeed, when compared with the glacial events in the northern hemisphere during the past 15,000 years, the Antarctic ice-sheet appears to have remained virtually unchanged.

3.4 Glacial eustatic changes

The term 'eustatic' is applied to those changes in sea-level that are simultaneous over the entire world. They can be caused by changes in the volume of the ocean basins, or by changes in the amount of water temporarily abstracted from the sea by glaciers and ice-sheets. If a warm climate turns into a cold glacial climate the precipitation over large areas changes from rain into snow. The snow becomes locked up in ice-sheets and does not form river-flow back to the oceans, thus resulting in a lowering of sea-level.

Decreases in sea-level during the formation of the Würm/Wisconsin ice-sheet are shown in Figure 5.11. A sea-level lowering of up to 120 m is suggested at 18,000 BP when the ice-sheets were at their greatest extent. The melting of the Antarctic ice-sheets would increase present world sea-levels by about 55 m. Thus short-term global sea-levels are controlled by the amount of ice in the world. Any change in ice-sheet volume leads immediately to a change in global sea-levels.

3.5 Ice surges

Glacial ice normally moves very slowly but periodically ice-sheets are known to become unstable and the ice to surge forward relatively quickly. Wilson (1964, 1966, 1969) has suggested that the Antarctic ice dome periodically becomes unstable and large masses of ice move forward into the Southern Ocean. One of the prerequisites of a surge is basal melting, allowing the ice to move relatively quickly. However, it appears that less than 10 per cent of the present Antarctic ice-cap is subject to melting processes near the ground, so it is not about to surge in the near future.

A massive surge of the Antarctic ice-sheet would cause a rapid rise in sea-level, which would then gradually decline as ice-sheets developed in Antarctica or elsewhere. Certainly large areas of the Southern Ocean would be covered by ice and the global albedo would be lowered, leading to a world-wide drop in temperature. The existence of a vast ice-shelf as postulated by the surge hypothesis probably would cause major changes in the position of the Antarctic convergence in the Southern Ocean. According to Hays (1967) such major disturbances did not occur during the Quaternary, since there were only minor migrations of the Antarctic convergence. Indeed, data from deep-sea cores seem to argue against massive surges from Antarctica with large associated ice-shelves in the Southern Ocean. However, they do not eliminate the possibility of smaller surges of individual drainage systems within the ice-sheet.

Flohn (1974) suggests that a surge spreading one-fourth or one-third of the present mass of the Antarctic ice dome more or less disintegrated into the Southern Ocean over a few decades or centuries, is not too unrealistic. Assuming an average thickness of 200 m for tabular icebergs, a nearly simultaneous outbreak of 6×10^6 km^3 would produce an ice-covered ocean area of 30×10^6 km^2. According to Flohn, the effect of an Antarctic surge of this size on the oceanic heat budget depends mainly on ice volume and temperature. Even more important is the effect on surface albedo, which

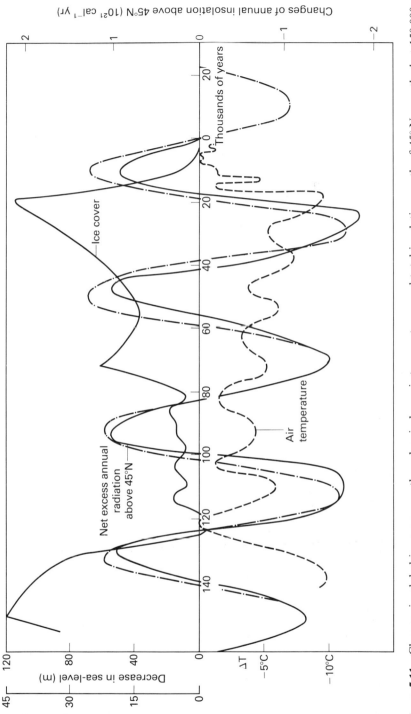

Figure 5.11 Changes in global ice cover, northern hemisphere air temperatures, and total insolation north of 45° N over the last 150,000 years. Total insolation plotted twice after Milankovitch and Vernekar (*after Mason, 1976*).

depends on the area covered by ice. Flohn concludes from climatological models that a surge causing a sea-ice area increase of 30×10^6 km^3 would cause a southern hemisphere temperature drop of 7–8 degree C.

3.6 The growth of the Laurentide Ice-Sheet

Flohn (1974) states that the key to the initiation of a northern glaciation (either complete or incomplete) is the summer climate of Labrador-Ungava, including Keewatin and probably Baffin Island. Andrews and Mahaffy (1976) consider that any major change toward a full glacial 'cycle' would be observed first across the 450,000 km^2 terrain of Baffin Island, the fifth largest island in the world. Evidence for this was first presented by Ives (1962) who commented on the extensive tracts of lichen-free uplands north of the present Barnes Ice-Cap and suggested that these reflected snowkill during a period when these uplands were mantled by thin snow-fields. Lichen studies and ^{14}C data by Andrews and Mahaffy indicate that these former snow-fields developed between 500 and 300 years ago and started to thin and retreat about 70 years ago. Furthermore, their mapping indicates that 300 years ago in the Little Ice Age the present glaciated area of 32,000 km^2 in central Baffin Island increased to an estimated 140,000 km^2, or by a factor of 4 in a matter of 100 to 200 years.

A knowledge of the topography and climate of these central areas in the eastern Canadian Arctic has led Ives (1957, 1958, 1962) to advance the concept of 'instantaneous glacierization'. Climatological support has been given to this theory by Lamb and Woodroffe (1970) and has led to the discussion of the rapid glacierization of large tracts of low and middle Canadian Arctic in the space of 1,000 to 5,000 years. This is in contrast to earlier views which estimated that not less than 15,000 years were required for the creation of a full-scale ice-sheet.

The rapid glacierization of large tracts in Arctic Canada suggests that the winter precipitation of 150 to 600 mm must have increased significantly. Lamb (1974) and Flohn (1974) have discussed the most probable circulation patterns that would lead to significant increases of solid precipitation over the Canadian Arctic. The pattern is dominated by a major meridional trough over Hudson Bay positioned to the west of the present-day upper cold trough. Such a pattern could lead to an increase in solid precipitation to 600–2,000 mm.

Andrews and Mahaffy (1976) have used a physically based ice-flow model to ascertain the rate at which the Laurentide Ice-Sheet might spread and thicken. Their results indicate that the fall in world sea-level that can be attributed to the growth of the Laurentide Ice-Sheet is slightly over 2 m after 5,000 years and nearly 20 m after 10,000 years. The ice-sheet after 10,000 years of development shows a major centre over Baffin Island/Foxe-Basin/ northern Keewatin and another larger centre occupying Labrador-Ungava. The broad pattern of ice distribution and inferred movement is in keeping with the scanty knowledge of flow patterns during the early Würm/Wisconsin glaciation. Andrews and Mahaffy consider that the hypothesis of 'instantaneous glacierization' is an acceptable hypothesis to explain the first few thousand years of ice-sheet development, but the continued growth of ice-sheets from the accumulation centres depends on the continued inputs of large amounts of mass. In their model they allowed a maximum net mass balance of 900 mm rainfall equivalent in the form of snow, a value three times that which is found over the Barnes Ice-Cap at the present time. Andrews *et al.* (1972, 1974) have concluded that extensive glacierization began on the plateaux of the eastern Canadian Arctic about 115,000 BP. Accumulation rates for the Camp Century ice-core indicate that between 120,000 and 70,000 BP the snowfall was equivalent to 600 to 1,200 mm of rainfall, representing an increase from the present by a factor of between 2 and 4. This particular problem is discussed again the next section.

4 The changing solar constant and world climate

Variations in solar insolation have occurred in the past because of changes in solar luminosity and also because of the Milankovitch mechanism. Wetherald and Manabe (1975) have shown that there is a marked positive feedback between the greenhouse effect of water vapour and surface temperature. This positive feedback increases the sensitivity of the surface temperature to changes in the solar constant by a factor of about 2.

For their investigation, Wetherald and Manabe (1975) used a simple mathematical model of the general atmospheric circulation. Since the model was relatively simple it should not be considered as a facsimile of the atmosphere, but nevertheless it does allow the basic properties of the atmosphere to be explored in some detail. Wetherald and Manabe used their model to explore in particular the effects of changes in the solar constant.

They found that the most dramatic effects were on the hydrological cycle. Thus it is seen from Table 5.1, that a 2 per cent increase in the solar constant

Table 5.1   Area-mean rate of precipitation for four different values of the constant in the general circulation model of Wetherald and Manabe, 1975 (relative magnitudes shown in brackets).

Solar constant ($1y \ min^{-1}$)		Rate of precipitation ($cm \ day^{-1}$)	
2·04	(1·02)	0·284	(1·10)
2·00	(1·00)	0·257	(1·00)
1·96	(0·98)	0·241	(0·94)
1·92	(0·96)	0·213	(0·83)

causes up to a 10 per cent increase in the rate of precipitation and therefore of evaporation. The change in precipitation rate resulting from changes in the solar constant is not uniform in all latitudes, but instead is particularly small in low latitudes and relatively large in middle and high latitudes.

Two basic mechanisms explain the marked intensification of the hydrological cycle with increasing solar constant. The temperature of the model atmosphere increases with increasing solar constant, leading in turn to a greater water vapour content. As the water vapour content of the atmosphere rises, so also does the absorption of infrared radiation from the surface, thus causing a decrease in the infrared cooling of the surface to space. The increase in atmospheric absorption and downward re-radiation more than compensates for the slight increase in surface temperature. Therefore the percentage increase in net radiation at the surface is slightly greater than the percentage increase in the solar constant, and so there is more energy available for evaporation, and the hydrological cycle is intensified. The mechanism is illustrated in Figure 5.12. The second mechanism concerns the Bowen ratio (ratio of sensible heat transfer to latent heat transfer), which for a wet surface in simple radiative equilibrium is a function of surface

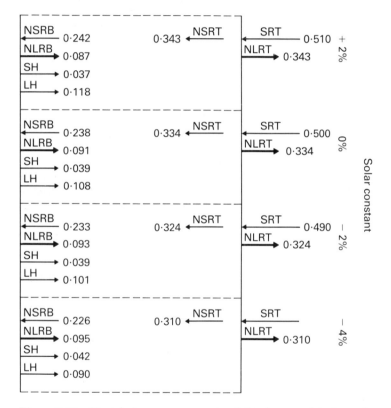

Figure 5.12   Heat balance components of the four model atmospheres considered by Wetherald and Manabe, 1975. In each case the percentage change in the solar constant is shown at the top of the column.

   SRT   incoming solar radiation at the top of the atmosphere
   NSRT   net downward solar radiation at the top of the atmosphere
   NLRT   net upward terrestrial radiation at the top of the atmosphere
   NSRB   net downward solar radiation at the earth's surface
   NLRB   net upward terrestrial radiation at the earth's surface
   SH   upward flux of sensible heat
   LH   upward flux of latent heat
Units are in ly min^{-1} (1 ly min^{-1} = 698 W m^{-2}).

temperature, as shown in Figure 3.3. Therefore the higher the temperature the greater the proportion of net radiation which goes into evaporation.

Wetherald and Manabe investigated the differences in zonal mean temperature for a 2 per cent rise in the solar constant, and found that net warming takes place. This warming is most pronounced in high latitudes because of the reduction of the area of snow cover with a large surface albedo. A feedback mechanism operates whereby the increase in the intensity of solar radiation raises the temperature of the earth's surface, reduces the areas of snow cover, and thus causes a further increase of surface temperature. Also the warming of high latitudes is essentially confined to a relatively shallow layer next to the earth's surface because of the suppression of vertical mixing in stable layers. Therefore most of the thermal energy involved affects the temperature of this shallow surface layer rather than being spread throughout the entire depth of the troposphere. Wetherald and Manabe comment that the tropospheric warming corresponding to the 2 per cent increase in the solar constant is comparable in magnitude to the tropospheric warming obtained for the doubling of the atmospheric CO_2 content.

Snowfall is also influenced by changes in the solar constant as shown in Figure 5.13. As the solar constant falls, both the amount of snowfall increases and the latitude of maximum fall decreases. This is mainly due to the sub-freezing polar temperatures spreading into the heavy precipitation zones of the westerlies. Figure 5.13 shows large increases in snow accumulation between latitudes 70° and 50°. Thus the model suggests that the snowfall in the higher middle latitudes will increase with a decreasing solar constant, while global precipitation (rain plus snow) totals actually decrease. This appears to offer an explanation for the increase in snowfall required for the growth of the middle-latitude ice-sheets when world precipitation totals were probably falling.

The model used in these calculations is a crude one and the changes predicted are only general indications of what might happen. If the changes in solar insolation were due to the Milankovitch mechanism, then they would not be world-wide and they would require a model with a rather more complex radiation input than the one described by Wetherald and Manabe. Nevertheless, their model probably does give some indication of what might happen when the solar radiation is decreasing in the northern hemisphere because of the Milankovitch mechanism.

Figure 5.11 plots against time the volume of the continental ice-sheets, the mean global surface temperature, and the variations in annual insolation received northwards of 45° N due to changes in the earth's orbit, using both Milankovitch's original calculations and the revised values of Vernekar (1972). The amplitude of the solar radiation variations is about 1 per cent of the total incident on the area. Clearly, there is a close correlation between the advances and recessions of the ice and the variations in solar insolation due to orbital changes.

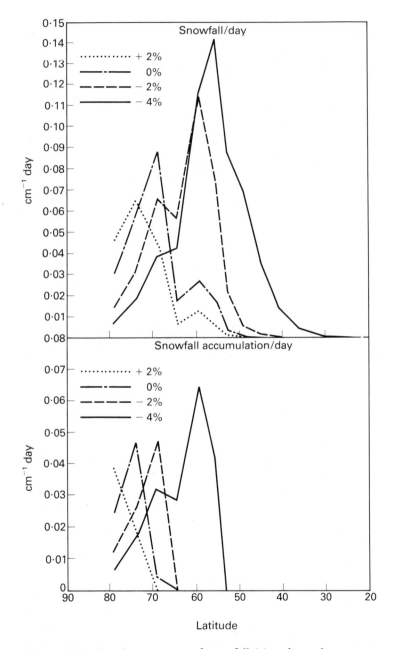

Figure 5.13　Zonal mean rates of snowfall (a) and zonal mean rates of snow accumulation (b), each for the indicated percentage values of the solar constant. Units are in cm day^{-1} (*after Wetherald and Manabe, 1975*).

5 Glacial climates

The Würm ice age reached its maximum about 18,000 year BP and represents the most dramatic change in the earth's surface for which widespread evidence is still available. Evidence of this paleoclimatic event is found in ocean and lake sediments, in records of soil structure and vegetative cover. In 1971 a consortium of scientists from a number of institutions was formed to study the history of global climate over the past million years, particularly the elements of that history recorded in deep-sea sediments. This study, known as the CLIMAP (Climate, long-range Investigation, Mapping and Prediction) project, is part of the United States National Science Foundation's International Decade of Ocean Exploration Program. One of CLIMAP's goals is to reconstruct the earth's surface at particular times in the past. These reconstructions can then serve as boundary conditions for atmospheric general circulation models.

The most detailed and highly developed models of climate are the general circulation models of the atmosphere. General circulation models do not address the problem of long-term climatic change directly, instead, they attempt to explain in dynamic terms the complex processes which, by balancing the earth's radiation budgets, maintain a particular climate in equilibrium. These models require that three boundary conditions be specified: the extent and elevation of permanent ice, the global pattern of sea surface temperature, and the continental geography. In addition, some information is required on albedo. This particular approach is illustrated for 18,000 BP using the work of the CLIMAP project members (1976) and Gates (1976a, 1976b).

5.1 The 18,000 BP continents

Continental geography 18,000 years ago can be approximated by today's configuration with a lowered sea-level, this being the result of the transfer of water from the oceans to the continental ice-caps. CLIMAP project members have made a conservative estimate of sea-level lowering of 85 m, but other authors have estimated lowerings of up to 130 m. The resulting continental outlines are shown in Figure 5.14. Reconstruction of the distributions of vegetation types during the last ice age (Figure 5.14) indicates that desert regions, steppes, grasslands, and outwash plains expanded at the expense of forests, yielding a slight increase in the albedo of land areas not covered by ice. The most striking feature of the world at 18,000 BP shown in Figure 5.14 is the northern hemisphere ice complex, consisting of land-based glaciers, marine-based ice-sheets, and either permanent pack ice or shelf ice. This complex stretched across North America, the polar seas, and parts of northern Eurasia, but nevertheless large arctic areas in Alaska and Siberia remained unglaciated. In the southern hemisphere the most striking difference was the winter extent of sea-ice, since changes in land-ice were small. Estimated ice-sheet contours are shown in Figure 5.14 and suggest that some of the land-based ice-sheets reached an approximate thickness of 3 km.

Figure 5.14 Sea surface temperatures, ice extent, ice elevation, and continental albedo for northern hemisphere summer (August) 18,000 years BP. Contour intervals are 1 degree C for isotherms and 500 m for ice elevation. Continental outlines represent a sea-level lowering of 85 m. In northern Siberia, dotted lines indicate a recently revised estimate of ice extent (*after CLIMAP Project Members, 1976*). Albedo values are given by the following key.

A: snow and ice; albedo over 40 per cent. Isolines show elevation of the ice-sheet above sea-level in metres.
B: sandy deserts, patchy snow, and snow covered dense coniferous forests, albedo between 30 and 39 per cent.
C: loess, steppes, and semi-deserts; albedo between 25 and 29 per cent.
D: savannas and dry grasslands; albedo between 20 and 24 per cent.
E: forested and thickly vegetated land; albedo below 20 per cent (mostly 15 to 18 per cent).
F: ice-free ocean and lakes, with isolines of sea surface temperature (°C); albedo below 10 per cent.

5.2 The 18,000 BP sea surface

One of the most important conclusions to be drawn from Figure 5.14 is that the ice-age sea surface temperature changes were not in general very large, since the average anomaly over the entire ocean was only −2·3 degree C. The geographic distribution of sea surface temperature anomalies is shown in Table 5.2. Additional insights into the spatial pattern of climatic change can be gained from a latitudinal plot of zonally averaged sea surface temperature anomalies as shown in Figure 5.15. On this plot the changes at

Table 5.2 Geographic distribution of the August ice age sea surface temperature anomaly, Δ*T*(degree C). Values are averages weighted by area, excluding regions covered by sea-ice (*after CLIMAP Project Members, 1976*).

Area	*T*(degree C)
Northern hemisphere	
Atlantic	− 3·8
Pacific	− 2·3
Indian	− 0·8
Average	− 2·6
Southern hemisphere	
Atlantic	− 1·7
Pacific	− 2·6
Indian	− 1·3
Average	− 2·0
Global ocean	− 2·3

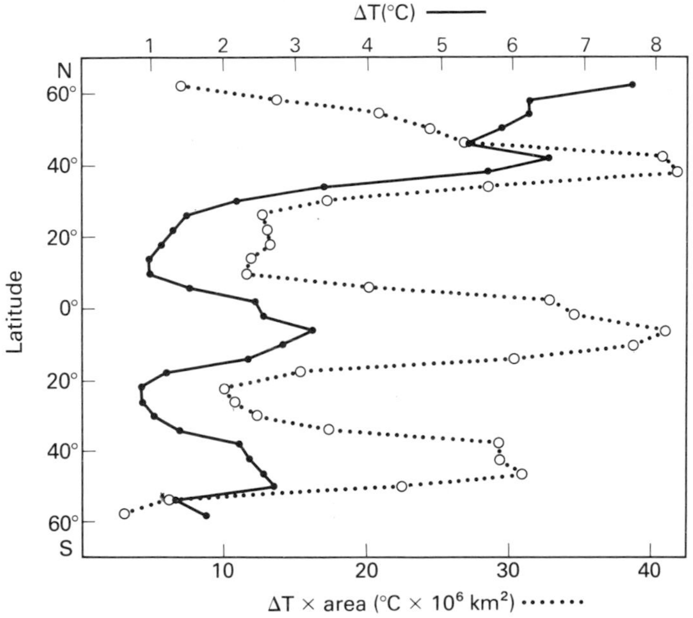

Figure 5.15 Average sea surface temperature difference between a modern August and August 18,000 BP (*after CLIMAP Project Members, 1976*).

high northern latitudes appear to dominate the pattern, but if this curve is weighted according to the area of ice-free ocean along discrete latitudinal bands, then the maximum effects occur at about 38° N, 6° S and 46° S and are roughly of the same magnitude. This suggests a marked steepening of thermal gradients and a more energetic ocean circulation system, particularly in the North Atlantic and Antarctic. Certainly surface transport was increased and probably a more rapid turnover of surface and intermediate waters was affected. Inferences about the structure of deeper waters are more tenuous.

A summary of area-averaged surface conditions for July 18,000 BP and present July are shown in Table 5.3. It should be noted that on a global basis, the average ice age sea surface temperature was 1·0 degree C below that of today's oceans in July, but if only the open ocean locations *common* to both cases are used the average ice age sea surface temperature cooling is 2·3 degree C. A maximum cooling of 17·2 degree C occurred in the western North Atlantic near 42° N, 60° W, while in the equatorial Pacific the ice age ocean was 7·1 degree C colder near 6° S, 150° W.

Table 5.3   Summary of area-averaged surface boundary conditions for ice age (18,000 BP) and present July (*after Gates, 1976a*).

	Ice age average		Present average	
Variable	Northern hemisphere	Southern hemisphere	Northern hemisphere	Southern hemisphere
Surface albedo	0·20	0·24	0·12	0·16
Sea surface temperature* (°C)	22·2	15·8	23·0	16·9
Ice-free water area† (10^6 km²)	129·1	166·0	142·7	186·5
Sea-ice area (10^6 km²)	9·7	34·5	10·4	19·8
Ice-sheet area‡ (10^6 km²)	31·7	17·2	4·3	13·1
Bare land area (10^6 km²)	84·5	37·3	97·6	35·6

* Averaged over ice-free ocean areas in each case.
† Including ice-free lakes.
‡ Including snow-covered land.

5.3 Simulated climatic conditions at July 18,000 BP

The global distribution of 18,000 BP July climate has been simulated by Gates (1976a, 1976b) with a two-level atmospheric general circulation model using the surface boundary conditions of sea surface temperature, ice-sheet topography and surface albedo assembled by CLIMAP. At the time of writing this is the best simulation available and the results are therefore discussed in some detail. The model probably gives a reasonable picture of the large-scale climate of the ice age.

5.3.1. Surface air temperature

July surface air temperature as simulated by Gates is shown in Figure 5.16. The temperature is that of the air immediately above the earth's surface, and is closely tied to the surface temperature itself. While ice age cooling is largest over the ice-sheets themselves, local temperature differences (present minus ice age) as large as 10 degree C are found over extensive unglaciated regions of North America, Europe and Asia.

5.3.2 Sea-level pressure

While most of the present-day features are also present in the ice age sea-level pressure distribution shown in Figure 5.17, there are a number of significant

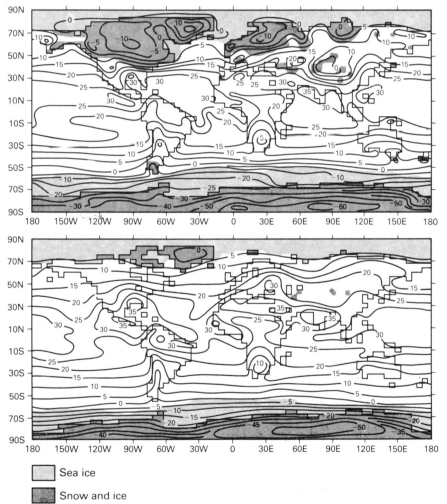

Figure 5.16   The surface air temperature (°C) simulated for the ice age July (above) and present July (below) (*after Gates, 1976b*).

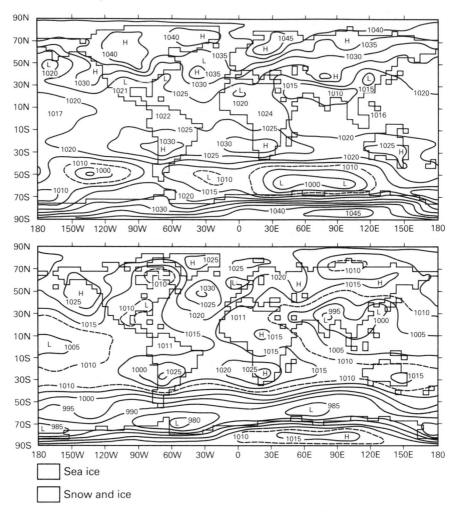

Figure 5.17   The sea-level pressure (mbar) simulated for the ice age July (above) and present July (below), referenced in each case to the appropriate sea-level (*after Gates, 1976b*).

differences. Aside from an average 12·7 mbar rise of sea-level pressure relative to the present, due largely to the 85 m lowering of ice age sea-level, the most dramatic ice age change from the present July pressure pattern is the introduction of prominent anticyclones over each of the major continental ice-sheets of North America, Europe and Siberia in response to the relatively cold air near the surface. The anticyclonic flow over the Laurentide ice-sheet in particular serves to introduce relatively cold air from the western North Atlantic into the North American continent to the south of the ice-sheet itself, and it is this flow which is responsible for much of the 10–15 degree C cooling of the surface air found over much of unglaciated North America.

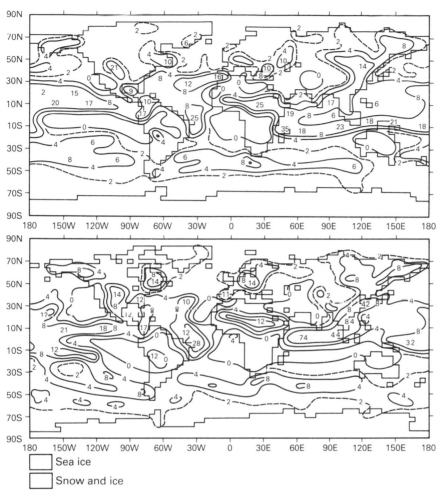

Figure 5.18   The precipitation rate (mm day^{-1}) simulated for the ice-age July (above) and present July (below). The isopleths are drawn every 2 mm day^{-1} except in regions of intense precipitation where only the local maximum is given (*after Gates, 1976b*).

Over Europe a similar alteration of surface air flow is simulated with the generally south-westerly flow of today's July replaced by an ice age circulation from the north and north-east. As in North America, this introduces air which is about 10 degree C colder than at present into the region south of the Scandinavian ice-sheet. In the tropics the most evident change of ice age sea-level pressure is the slight westward displacement and overall weakening of the low-pressure system associated with the summer monsoon over southern Asia. Gates comments that while the simulated ice age pressure distribution is in general difficult to verify, the pressure distribution over the North Atlantic bears some resemblance to that reconstructed from paleoclimatic evidence by Lamb (1971) and Lamb and Woodroffe (1970).

5.3.3 *Precipitation*

Global distributions of the daily precipitation rate for both the ice age and present July are shown in Figure 5.18. On a globally-averaged basis, the simulated total precipitation is approximately 14 per cent less in the ice age than at present, with most of the reduction occurring in the northern hemisphere as a result of reduced convective activity.

In detail, the most notable changes from the present are the reduction of ice age precipitation over the Scandinavian and Laurentide ice-sheets, and the reduction of the monsoonal rainfall over south-east Asia. In the latter region, the heavy precipitation simulated for the present July over the equator in the Indian Ocean is shifted southward to about 10° S in the ice age. Associated with this change is an increase of the precipitation along the eastern coast of central Africa.

Data presented in Table 5.4 show that the overall reduction of the ice

Table 5.4   Area-averaged precipitation at July 18,000 BP according to Gates, 1976b.

	Ice age		Present		Global difference (ice age minus present)
	Northern hemisphere	Southern hemisphere	Northern hemisphere	Southern hemisphere	
Precipitation rate (mm day^{-1})					
Large-scale	0·31	0·90	0·37	0·81	+0·01
Convective	4·31	2·24	5·39	2·41	−0·62
Total	4·61	3·14	5·75	3·22	−0·61

Ice age differences with respect to present July	Northern hemisphere	Southern hemisphere
Ocean	−0·60	+0·04
Sea-ice	+0·06	−0·95
Land	−1·91	+0·42
Ice-sheet	−1·75	−0·47
All surfaces	−1·14	−0·08

age precipitation relative to the present is primarily due to an approximate 20 per cent reduction in the rainfall associated with convective activity in the northern hemisphere. In the southern hemisphere, on the other hand, the convective precipitation was only slightly reduced in the ice age, while the precipitation associated with large-scale vertical motion was even greater in the ice age than at present. The geographical partitioning of these precipitation changes is shown in Table 5.4. It is seen that the total precipitation in the northern hemisphere decreased much more over the bare land and ice-sheets than over the oceans, while the ice age precipitation is simulated to have been greater than at present over the largely unglaciated continents of the southern hemisphere.

5.3.4 *Evaporation*

In general the simulated ice age evaporation is less than that at present in accordance with the generally lower ocean surface temperatures. The globally-averaged reduction during the ice age amounts to about 15 per cent, most of which occurs in the southern hemisphere. There is an overall decrease of the intensity of the global hydrological cycle as measured by the difference between evaporation and precipitation, although on a regional basis it has been weakened over the ocean in the southern hemisphere while strengthening markedly over the land in the northern hemisphere.

5.4 The British climate during the last glaciation

Despite some uncertainty about the precise dates, evidence suggests according to Lamb (1977) that at some time between about 70,000 to 60,000 BP ice-sheets were quickly established over northern and north-western Europe, probably including parts of the British Isles, and over extensive areas of North America. At about 50,000 BP the ice-sheets reached a secondary maximum, and then retreated only to advance again to reach their maximum extent at about 18,000 BP. These changes are illustrated in Figure 5.11. Between the climaxes of low temperature and ice cover in the early and late glacial there were many thousands of years of variable, rather more temperate conditions, although these were generally colder than now. For much of the time the seas continued cold, and the great ice-sheets, though somewhat reduced, remained in existence.

Coope (1975a, 1975b) has estimated July temperatures in Britain during the Würm/Wisconsin glaciation using fossil assemblages of insect fauna (Coleoptera). The main results of his investigations are shown in Figure 5.19. Apart from several interstadial periods, average lowland July temperatures were of the order of 9 or 10 °C, falling to 8 °C at the maximum expansion of the ice-sheets. After the Chelford interstadial about 60,000 BP, when pine, spruce and birch grew in the English Midlands, there was no forest flora in England for about 50,000 years, until the return of a predominantly birch forest at the time of the Allerød oscillation. During the Chelford interstadial the climatic regime indicated by the insect assemblages is similar to that of south central Finland at the present day, that is with average July temperatures of about 15 °C, a degree or two lower than those in the Cheshire Plain today. Also climatic continentality was somewhat greater than in present-day England.

The fossil Coleoptera indicate that during the Upton Warren interstadial there was only one major episode of climatic amelioration, which at its thermal maximum must have involved average July temperatures of about 18 °C, that is 1 or 2 degree C above those of southern and central England at the present day. The fauna as a whole suggests a degree of continentality no greater than that of central Europe today. A peculiarity of the Coleoptera, which marks them out as different from the present-day faunas of central Europe, is the total lack of any species that are dependent on trees, either directly as a source of food or indirectly as a source of specialized habits. This absence of trees is supported by palynological studies which show little

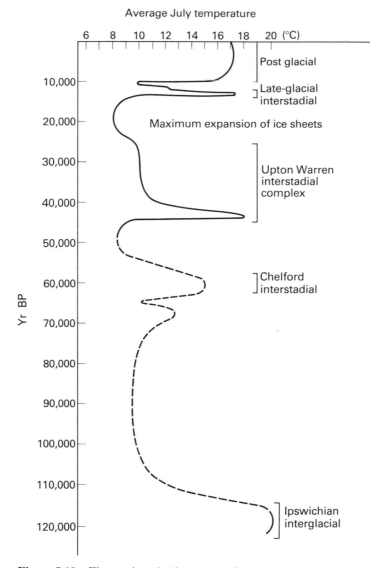

Figure 5.19   Fluctuations in the average July temperature in lowland Britain since the Eemian (Ipswichian) interglacial (*after Coope, 1975*).

tree pollen other than that which has been blown in from far afield or re-deposited pollen eroded out of earlier strata. Coope (1975b) has given a likely explanation for the absence of trees at this time based on the historical setting of this short but intense climatic amelioration. It was preceded by a period of intense cold, quite adequate to eliminate the forest from western Europe and reduce the landscape to barren tundra. The onset of the climatic amelioration was so sudden and its duration so brief that the forest may simply not have had time enough to spread from its refuges in the south. Heavy grazing along the forest limit by herds of bison and reindeer might also have inhibited regeneration and thus hindered its spread northwards.

There was a gradual deterioration of climate from the warm interlude of the Upton Warren interstadial until the period of maximum expansion of the ice-sheets in Britain which occurred about 18,000 BP. For much of the time the average July temperatures must have been at or lower than 10 °C, also the climate of the times was very much more continental than anywhere in Europe at the present day. Coope suggests that the climatic continentality must have entailed the development in Britain of exceedingly cold winters with average February temperatures below − 20 °C. Towards the end of the period there is some evidence of increased oceanicity of climate, probably providing the necessary precipitation for the growth of the glaciers at this time.

During the period between the warm interlude of the Upton Warren interstadial and about 18,000 BP, the climate of Britain must have been very cold, because permafrost and related phenomena such as ice-wedges were able to develop. Williams (1975) comments that features indicative of former permafrost have been observed in many parts of Britain and it is clear that at one time or another permafrost must have existed throughout most, if not all, of the country. Ice-wedges, unlike other bodies of ice associated with permafrost, actually require somewhat colder conditions than permafrost in order to develop, and are therefore useful indicators of past temperatures. Ice-wedges grow by repeated cracking and infilling with ice, usually in areas where the permafrost is subjected to severe cooling. Williams (1975) concludes from the evidence of ice-wedges in Britain, that mean annual temperatures were of the order of − 8 or − 10 °C during the period leading up to the maximum extent of ice. Winter temperatures were probably very low, reaching − 25 °C, with sea-ice off the coasts of Britain.

Precipitation in glacial times is even harder to estimate than the temperatures. The presence of ice-wedges in Britain is an indication of low winter snowfall during the period when they were growing. Wedges will not crack, and therefore will not grow if there is a thick cover of snow over the ground. The insulating effect of snow does not increase uniformly with increasing depth, since once a certain thickness of snow is reached practically all transfer of heat is stopped. The critical thickness appears to be between 15 and 25 cm.

It seems safe to conclude that winter snowfall was light while ice-wedges were growing in Britain during the period before the maximum expansion of the ice-sheets. Williams (1975) states that if an upper limit of 25 cm of snow is accepted, then winter precipitation must have been equivalent to about 100 mm of rainfall, considerably drier than at present. About 60 per cent of the total precipitation in many Arctic areas falls during the summer months, and if there was a similar seasonal variation in periglacial Britain, the annual precipitation in areas with ice-wedges could not have exceeded about 250 mm of rainfall.

A simulated water balance for a lowland British site during the cold dry period before the maximum expansion of the ice-sheets is shown in Table 5.5. The values are for illustration only and do not represent actual obser-

Table 5.5 Simulated water balance of a site in lowland Britain during a dry period in the Würm/Wisconsin glaciation. (*Except for temperature, all values are in rainfall equivalents. Temperatures after Williams, 1975.*)

Month	Precipi- tation (mm)	Snow storage (mm)	Direct evapor- ation of snow (mm)	Actual evapor- ation (mm)	Run-off (mm)	Soil moisture (end of month) (mm)	Temper- ature (°C)
J	19·8	66·5	0		0	frozen	−24
F	14·7	78·2	3		0	frozen	−25
M	12·0	81·2	9		0	frozen	−21
A	20·6	83·8	18		0	frozen	−10
M	24·0	0		50	10	47·8	0
J	18·7	0		66·5	0	0	+5
J	27·6	0		27·6	0	0	+10
A	28·6	0		28·6	0	0	+9
S	24·0	0		20	0	4	+5
O	17·9	12·9	5		0	frozen	−4
N	20·0	28·9	4		0	frozen	−16
D	17·8	46·7	0		0	frozen	−24
Total	245·7						

vations. Winter snowfall is kept low to agree with Williams' estimates, and there is no winter run-off because the ground is frozen. In May temperatures rise above 0 °C and the snow melts both saturating the soil and causing a brief period of direct run-off, which could be intense for a short time. Summer evaporation exceeds rainfall, and the soil soon becomes dry giving a summer of extreme aridity. In October the first snows will accumulate on ground which is rather dry. Changing the values of the climatic elements by relatively small amounts does not modify the general picture, which is one of late spring run-off and general summer aridity.

The final advance of the ice-sheets somewhat before 20,000 BP, was probably caused by an increase in precipitation together with slightly lower summer temperatures. Gates (1976a, 1976b) suggests a July rainfall at about 18,000 BP over Britain of slightly less than 60 mm, indicating that precipitation values may have been twice those used in Table 5.5, and not too far short of present-day values. Because direct evaporation from snow would remain constant, doubling the winter precipitation would more than double the water held in snow storage at the end of the winter in April shown in Table 5.5, values probably exceeding 205 mm. Such an extensive snow cover would probably last until July before it vanished and would give rise to extensive spring and summer run-off as it melted. Temperature falls with increasing elevation, so the summer temperatures on the uplands and the ice domes would have been at or below 0 °C, allowing an annual accumulation of snow. Thus a substantial increase of precipitation on the ice domes would cause them to advance across the lowlands. Since the glaciers had already abandoned almost all of lowland Britain before the rise in temperature at the start of the next interstadial episode, it seems likely that precipitation deficiency played a significant role in the waning of these ice-sheets.

References

ANDREWS, J. T., BARRY, R. G., BRADLEY, R. S., MILLER, G. H. and WILLIAMS, L. D. 1972: Past and present glaciological responses to climate in eastern Baffin Island. *Quaternary Research* **2**, 303–14.

ANDREWS, J. T., FUNDER, S., HJORT, C. and IMBRIE, J. 1974: Comparison of the glacial chronology of eastern Baffin Island, east Greenland, and the Camp Century accumulation record. *Geology* **2**, 355–8.

ANDREWS, J. T. and MAHAFFY, M. A. W. 1976: Growth rate of the Laurentide ice sheet and sea level lowering (with emphasis on the 115,000 B.P. sea level low). *Quaternary Research* **6**, 167–83.

BJERKNES, J. 1969: Atmospheric teleconnections from the equatorial Pacific. *Monthly Weather Review* **97**, 163–72.

CLIMAP Project members 1976: The surface of the ice-age earth. *Science* **191**, 1131–7.

COOPE, G. R. 1975a: Climatic fluctuations in northwest Europe since the last interglacial, indicated by formal assemblages of Coleoptera. In WRIGHT, A. E. and MOSELEY, F., editors, *Ice-ages: ancient and modern. Geological Journal*, special issue **6**, 153–68.

—— 1975b: Mid-Weichselian climatic changes in western Europe, reinterpreted from Coleopteran assemblages. In *Quaternary studies: selected papers from IX INQUA Congress Christchurch, New Zealand, 1973.* Wellington: Royal Society of New Zealand.

DAMON, P. E. and KUNEN, S. M. 1976: Global cooling? *Science* **193**, 447–53.

DANSGAARD, W., JOHNSON, S. J., CLAUSEN, H. B. and LANGWAY, C. C. JR, 1971: Climates record revealed by the Camp Century ice core. In TUREKIAN, K. K., editor, *The late Cenozoic glacial ages*, 37–56. New Haven: Yale University Press.

—— 1975: Speculations about the next glaciation. *Quaternary Research* **2**, 396–8.

DIETZ, R. S. and HOLDEN, J. C. 1970: Reconstruction of Pangaea: breakup and dispersion of continents, Permian to present. *Journal Geophysical Research* **75**, 4939–56.

FLOHN, H. 1974: Background of a geophysical model of the initiation of the next glaciation. *Quaternary Research* **4**, 385–404.

—— 1975: History and intransitivity of climate. In GARP, *The physical basis of climate and climate modelling*, 106–18. Geneva: WMO.

GATES, W. L. 1976a: Modelling the ice-age climate. *Science* **191**, 1138–44.

—— 1976b: The numerical simulation of ice-age climate with a global general circulation model. *Journal Atmospheric Sciences* **33**, 1844–73.

HAYS, J. D. 1967: Quaternary sediments of the Antarctic Ocean. In SEARS, M., editor, *Progress in Oceanography* **4**, 117–31. Oxford: Pergamon Press.

HAYS, J. D., SAITO, T., OPDYKE, N. D. and BURCKLE, L. H. 1969: Pliocene–Pleistocene sediments of the equatorial Pacific, their paleomagnetic, biostratigraphic and climatic record. *Bulletin Geological Society of America* **80**, 1481–514.

HAYS, J. D., IMBRIE, J. and SHACKLETON, N. J. 1976: Variations in the earth's orbit: pacemaker of the ice ages. *Science* **194**, 1121–32.

IVES, J. D. 1957: Glaciation of the Torngat Mountains, northern Labrador. *Arctic* **10**, 67–87.

1958: Glacial geomorphology of the Torngat Mountains, northern Labrador. *Geographical Bulletin* **17**, 47–75.

1962: Indications of recent extensive glacierization in northcentral Baffin Island, **NWT**. *Journal Glaciology* **4**, 197–205.

KENNETT, J. P. 1977: Cenozoic evolution of Antarctic glaciation, the circum-Antarctic Ocean, and their impact on global paleoceanography. *Journal Geophysical Research* **82**, 3843–60.

KUKLA, G. J. 1970: Correlations between loesses and deep-sea sediments. *Geol. Foren, Stockholm Forh* **92**, 148–80.

1975: Missing link between Milankovitch and climate. *Nature* **253**, 600–3.

KUKLA, G. J. and KUKLA, H. J. 1974: Increased surface albedo in the northern hemisphere. *Science* **183**, 709–14.

KUTZBACH, J. R. and BRYSON, R. E. 1974: *Variance spectrum of Holocene climatic fluctuations in the North Atlantic sector*. Madison: Department of Meteorology, University of Wisconsin.

1975: Variance spectrum of Holocene climatic fluctuations in the North Atlantic sector. In *Proceedings WMO/IAMAP symposium on long-term climatic fluctuations*, 97–104. Geneva: WMO.

LA MARCHE, V. C. JR. 1974: Paleoclimatic inferences from long tree-ring records. *Science* **183**, 1043–8.

LAMB, H. H. 1966: Climate in the 1960's. *Geographical Journal* **132**, 183–212.

1969: Climatic fluctuations. In FLOHN, H., editor, *World survey of climatology* **2**, *General climatology*, 173–249. New York: Elsevier.

1971: Climates and circulation regimes developed over the northern hemisphere during and since the last ice age. *Palaeogeography, Paleoclimatology, Paleoecology* **10**, 125–62.

1974: Atmospheric circulation during the onset and maximum development of the Wisconsin/Würm ice age. In HERMAN, Y., editor, *Marine geology and oceanography of the Arctic seas*, 349–58. New York: Springer-Verlag.

1977: *Climate: present, past and future* **2**, *Climatic history and the future*. London: Methuen and New York: Barnes and Noble.

LAMB, H. H. and WOODROFFE, A. 1970: Atmospheric circulation during the last ice age. *Quaternary Research* **1**, 29–58.

LORENZ, E. N. 1968: Climatic determinism. In MITCHELL, J. M. JR, editor, *Causes of climatic change*. Meteorological Monograph **8**, 1–3. Boston: American Meteorological Society.

MASON, B. J. 1976: Towards the understanding and prediction of climatic variations. *Quarterly Journal Royal Meteorological Society* **102**, 473–98.

PAINTING, D. J. 1977: *A study of some aspects of the climate of the northern hemisphere in recent years*. Scientific Paper 35. London: Meteorological Office.

SANDERSON, R. M. 1975: Changes in the area of Arctic sea ice 1966 to 1974. *Meteorological Magazine* **104**, 313–23.

SHACKLETON, N. J. and OPDYKE, N. D. 1973: Oxygen isotope and paleomagnetic stratigraphy of equatorial Pacific core V28–238: oxygen isotope temperatures and ice volumes on a 10^5 and 10^6 year scale. *Quaternary Research* **3**, 39–55.

UNTERSTEINER, N. 1975: Sea ice and ice sheets and their role in climatic variations. In GARP, *The physical basis of climate and climate model-ling* 206–24. Geneva: WMO.

US NATIONAL ACADEMY OF SCIENCES 1975: *Understanding climatic change, a program for action.* Washington, DC.

VERNEKAR, A. D. 1972: *Long-period global variations of incoming solar radiation.* Meteorological Monographs **12**. Boston: American Meteorological Society.

WETHERALD, R. T. and MANABE, S. 1975: The effects of changing the solar constant on the climate of a general circulation model. *Journal Atmospheric Sciences* **32**, 2044–59.

WILLIAMS, R. B. G. 1975: The British climate during the last glaciation: an interpretation based on periglacial phenomena. In WRIGHT, A. E. and MOSELEY, F., editors, *Ice ages: ancient and modern. Geological Journal,* Special issue **6**, 95–120.

WILSON, A. T. 1964: Origin of ice ages: an ice shelf theory for pleistocene glaciation. *Nature* **201**, 147–9.

1966: Variation in solar insolation to the south polar region as a trigger which induces instability in the Antarctic ice-sheet. *Nature* **210**, 477–8.

1969: The climatic effects of large-scale surges of ice sheets. *Canadian Journal Earth Sciences* **6**, 911–18.

6
The Climatic Future: Climatic Models and Trends

1 Climatic models

The climate of the earth may be considered in terms of forcing functions and response functions (Lettau 1969). The primary forcing function is solar radiation, and its operation is seen in the various natural cycles ranging from annual to diurnal, and in particular in local air and ground temperatures. The physical relationship between forcing and response cycles follows from the transformations of absorbed solar energy, which generates the fluxes that constitute the energy balance of the earth's surface. This was described in detail in Chapter 3. The other major forcing function is precipitation, since without precipitation there could be neither soil moisture nor evapotranspiration nor run-off.

Simple climatic models were discussed in Chapter 3, particularly with reference to annual periods. The approach will now be further illustrated by the discussion of a complex climatological model of the central Pennines, northern England. A flow diagram of the model is shown in Figure 6.1. The parameterized forms of the various energy and water exchanges are simplified representations of highly complex physical processes and assume some averaging in both space and time. The model follows the traditional view of water movement in a catchment as being part of a cascading system with inputs, outputs and several stores. The main stores are the canopy interception and surface water store, the soil moisture store and the groundwater store. The main inputs are precipitation and climatological information to determine evapotranspiration. Evapotranspiration was determined using Penman's method, which was described in Chapter 3, the most important climatological input being solar radiation. The main outputs are evapotranspiration and run-off. The run-off can be divided into two basic components which are a direct run-off component and a groundwater component. The direct run-off component originates directly from water held in the soil and on the soil surface, and responds rapidly to precipitation. The groundwater component comes from water held in the deeper rocks and does not respond directly to precipitation.

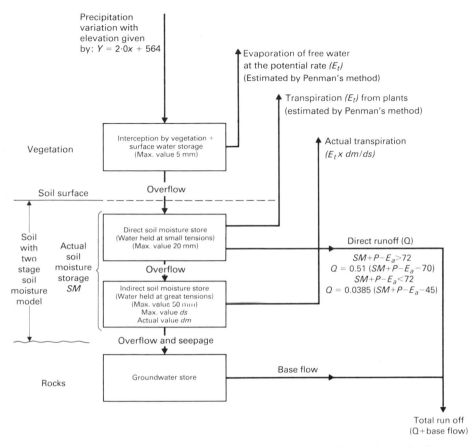

Figure 6.1 Climatological model of the central Pennines. Stores shown by boxes and transfers by arrows. Equations assume averaging in both space and time. Rainfall, Y, P; 8 km mean elevation, x; total soil moisture storage, SM; maximum value of indirect soil moisture storage, d_s; actual value of indirect soil moisture storage, d_m; E_t and E_a potential and actual evapotranspiration; direct run-off, Q.

1.1 Global radiation

A variety of forms of parameterization may be used to obtain the incoming global radiation. Penman (1948) estimated global radiation (R_G) from:

$$R_G = F(0.18 + 0.55\, n/N)$$

where F is the daily solar radiation incident on a horizontal surface in the absence of the earth's atmosphere, and n/N is the ratio of actual to possible hours of bright sunshine.

Comparison with global radiation measured on the edge of the Pennines near Harrogate suggested that while this relationship gave good estimates for most of the year it slightly overestimated global radiation in the summer. This was mainly because of the greater precipitable water and aerosol content of the atmosphere in summer and as a consequence a greater absorption of global radiation at this time of year.

It is possible to predict the daily march of global radiation (R_G) from equations of the type:

$$R_G = (a + b \sin \theta)S \sin \theta$$

where a and b are constant, θ is the solar angle (angle between horizon and sun), and S is the solar constant.

For the Pennines $a = 0.477$ and $b = 0.363$. During June, July and August the equation becomes:

$$R_G = (-0.021 + 0.718 \sin \theta)S$$

Such equations allow the global radiation to be estimated for any given time on a cloudless day and also the total amount of global radiation for a cloudless day to be estimated. Cloud may be taken into account by multiplying R_G by a factor F, thus

$$F = (1 - 0.4C_h)(1 - 0.7C_m)(1 - 0.7C_l)$$

where C_h, C_m, C_l are the percentages of the sky covered by high, medium and low cloud respectively. Thus it is possible to estimate global radiation under any given cloud conditions.

If the number of hours of sunshine per day is known, it is possible to estimate daily global radiation totals from

$$R_G = I(0.25 + 0.75 \, n/N)$$

where I is the daily global radiation under cloudless skies and may be obtained from the equation given above for R_G and n/N is the ratio of actual to possible hours of bright sunshine.

1.2 Evapotranspiration

Evapotranspiration was calculated in the model using Penman's method, which is described in Chapter 3. Evapotranspiration was assumed to be at the potential rate provided that the vegetation was wet from rainfall, or provided that the soil was moist. According to Hounan *et al.* (1975) the following concepts describing the extraction of water from the soil by plants have been used in the last 30 years.

(i) Evapotranspiration independent of soil moisture. A very simple approach but not reliable.

(ii) Evapotranspiration at potential rate, but reducing in a very dry soil. A variation of this method is used by Penman (1963), who assumes that plants transpire all water from the 'root zone', plus 25 mm of water drawn from below this level, at the potential rate. When this supply is exhausted, transpiration is reduced to one-tenth of the potential rate.

(iii) Evapotranspiration linearly proportional to soil water. This model has been used by a number of workers, notably Thornthwaite and Mather (1955), under conditions of high evaporative demand.

(iv) Evapotranspiration decreasing exponentially. Numerous models of this type have been developed showing a wide variation according to vegetation and soil type and possibly also season as root density and canopy change. Some models incorporate periods, when water content is high, during which evapotranspiration is at the potential rate; thereafter exponential decay takes over.

(v) Multilevel models. In its simplest form water storage is separated into two levels, the 'upper level', which contains most of the plant roots, and the 'lower level', which contains fewer roots. Water in the 'upper' zone is depleted at the potential rate and any deficiency in this zone must be satisfied by rainfall before re-charge of the lower zone commences. Depletion from the lower zone occurs only when there is no water available in the upper zone, the rate of evapotranspiration being proportional to the amount of water available in the lower zone.

A two-stage model of the last type was selected for use in the model. It was considered though that the stages do not necessarily correspond to physical layers in the soil but rather to physical states of the soil during the drying process. Evapotranspiration was assumed to take place from the first stage at the potential rate, and it was also assumed that precipitation initially enters the first stage. When the first stage was dry, evapotranspiration (E_a) took place from the second stage such that:

$$E_a = E_t \frac{dm}{ds}$$

where E_t is the potential evapotranspiration; ds is the maximum available water content of the second stage; and dm is the actual available water content of the second stage.

When the first stage becomes saturated, excess water not forming evapotranspiration or run-off was transferred into the second-stage storage. After rainfall the vegetation canopy will normally be wet and the surface acts as a saturated one regardless of the soil-moisture state, so in the model intercepted and surface water was included in the first stage for the purpose of calculating evapotranspiration. It was assumed that direct run-off originated mostly from the first stage of the soil-moisture store. Water in the second stage of the soil-moisture store did not normally form run-off, since this stage represented water which was tightly bound in the soil was compared with that in the first stage. Water transfer to the groundwater store (Figure 6.1) in the model only took place when there was a transfer of water from the first stage to a saturated second stage, but under actual saturated soil conditions there is probably an extremely slow seepage of water out of the soil into the deeper rocks.

The calibration of a model of this type is discussed in a paper by Lockwood and Venkatasawmy (1975). Priestley and Taylor (1972) using a similar model to the one just described, found a maximum value of 50 mm rainfall equivalent for their second-stage storage at several widely different sites. Numerous calculations with several years of data in the case of the Pennine model led to a maximum second-stage storage of 50 mm, and a maximum value of the final-stage storage of 25 mm rainfall equivalent including 5 mm for surface and vegetation storage. Thus the maximum available water content of typical Pennine Millstone Grit grassland is 75 mm if surface and vegetation storage is included. Direct observations of soil moisture by gravimetric methods suggested a maximum available moisture content in the grass rooting zone of about 60 mm.

Grindley (1972) has listed the maximum available soil moisture contents for a variety of crops in Britain and his values are listed in Table 3.1. It is seen that the maximum available soil-moisture content is very much a function of crop type and that the particular soil type is not of great importance. The greatest available moisture contents are associated with deep-rooting plants such as wheat or orchard trees, while the least are associated with poor grass crops. Grindley suggests a value of 50 mm for rough grazing, so 70 mm for a mixture of rough grazing and permanent grass over the Pennines is a reasonable value for the maximum available water contents.

In practice the available soil-moisture content will vary slightly with soil type and more directly with relief. Fedoseev (1958) suggests the use of a 'wetness coefficient' representing the ratio of total moisture content in one metre of soil in a typical orographic element to that in a flat field. Typical values are shown in Table 6.1 together with estimated values of maximum

Table 6.1   Mean values of the 'wetness' coefficient for different types of orographic relief (*after Fedoseev, 1958*). (Values in brackets are appropriate for the given relief assuming a flat field value of 70 mm maximum available soil moisture.)

Hillock relief			
Flat surface between hillocks	1·00	(70 mm)
Top of hillock	0·46	(32 mm)
Southern slope	0·53	(37 mm)
North-western slope	0·71	(49 mm)
Hollow relief			
Flat watershed	1·00	(70 mm)
Southern slope	0·76	(53 mm)
Trough on southern slope	1·07	(75 mm)
Foot of southern slope	0·94	(66 mm)
Lowland meadow	1·41	(99 mm)

available soil moisture for the Pennines assuming a flat field value of 70 mm. Table 6.1 suggests grassland values of maximum available soil moisture for the Pennines ranging from approximately 100 mm for lowland meadows, 50 mm for many slopes and about 30 mm for hillock tops.

Estimates of evapotranspiration and soil moisture in a typical $8·1$ km^2 grassland Pennine catchment are shown in Figure 6.2. Precipitation and run-off were measured while potential evapotranspiration was estimated by Penman's method. The potential evapotranspiration follows very closely the annual march of global radiation, being at a maximum in summer and a minimum in winter. In winter soil-moisture values are near the upper limit, though the soil is not saturated all the time. During the summer evapo-

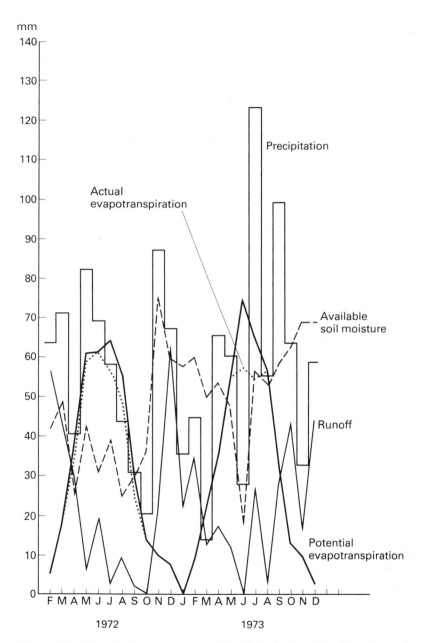

Figure 6.2 Crimple Beck Catchment, North Yorkshire. Monthly values of evapo-transpiration, run-off and soil moisture. All values are monthly totals except soil moisture which is the end of month value. Units: rainfall equivalents in mm (*after Lockwood and Venkatasawmy, 1975*).

transpiration exceeds rainfall and the soil dries causing the available soil moisture to reach low values. Under summer conditions the first-stage soil moisture store soon becomes exhausted and actual evaporation values fall below the potential rate.

1.3 Run-off

Under dry weather conditions river flow consists completely of water contributed by the aquifers bordering the river. Since groundwater forms most of the continuing long-term flow of the river, it is said to form the dry weather flow or base flow of the river. Since aquifers discharge decreasing amounts of water with time as the stored water decreases, the base flow slowly decays with time in an exponential manner. Base flow can be represented very nearly by

$$Q_t = Q_0 e^{-\alpha t}$$

where Q_0 is the discharge at the start of the period, Q_t is the discharge at the end of time t, α is a constant, and e is the base of natural logarithms.

A typical dry weather flow curve is illustrated in Figure 6.3 for the River Ure at Kilgram Bridge during the drought of 1976. It is seen that for the Pennine rivers the base flow is small and at times only equals about 0·05 mm rainfall equivalent per day.

When precipitation occurs which is sufficiently intense to produce surface or subsurface flow, water will start to move relatively quickly across the landscape and the river flow will rise rapidly. Significant surface or subsurface flow will occur when the rainfall intensity exceeds the infiltration capacity of the soil or when the surface layers of the soil become saturated. If it has been dry for some considerable period of time before the rainfall, then the river flow will consist completely of base flow, i.e. of groundwater which has seeped into the river. Once surface run-off reaches the river after the start of rainfall, the river flow will consist of both surface run-off and base flow, and as the river rises the surface run-off will become the dominant part of the river flow. Since seepage into and out of the aquifers is slow, base flow will remain steady or only increase slowly throughout the rain storm, and it may therefore be assumed that the short-term variations in river flow are largely due to surface run-off.

An examination of data from the Pennine catchments described by Lockwood and Venkatasawmy (1975), suggests that in the rather low intensity Pennine rainfall environment (usually less than 20 mm h^{-1}), a relationship exists between the total daily direct run-off and the available soil moisture. Figure 6.4 shows the results if total daily run-off minus estimated groundwater inflow, is plotted against the available soil moisture (SM) at the start of the day, plus the precipitation (P) and minus the actual evapotranspiration (E_a) during the day. Some scatter in the points is to be expected since there are uncertainties in the observations, and in particular the soil moisture estimates may be in error, while large rainfalls may have been incorrectly recorded. While complex curves are theoretically most likely, the data only justified the drawing of two straight lines. For values of total water content

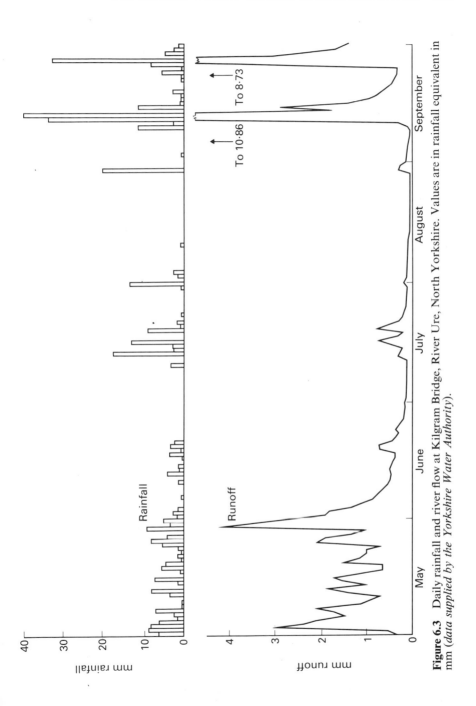

Figure 6.3 Daily rainfall and river flow at Kilgram Bridge, River Ure, North Yorkshire. Values are in rainfall equivalent in mm *(data supplied by the Yorkshire Water Authority)*.

(SM + P − E$_a$) below 72 mm, the simplest curve (curve A, Figure 6.4) fitting the data is

$$Q = 0.0385 \ (SM + P - E_a - 45)$$

while for values of total water content above 72 mm, the direct run-off (curve B, Figure 6.4) is

$$Q = 0.51 \ (SM + P - E_a - 70)$$

It now becomes clear that soil moisture is the main response function to the two forcing functions of precipitation and solar radiation. Soil moisture in turn determines the direct run-off and the actual evapotranspiration.

Figure 6.4 suggests that three main landscape states can be recognized, and these may be termed dry, moist and saturated. The dry state occurs with saturation levels (SM + P − E$_a$) up to about 45 mm, and approximately corresponds to the second stage of the soil moisture model described earlier. While the landscape remains in this state there is no direct run-off and the actual evaporation is below the full potential rate. Additional rainfall does not cause any significant run-off provided that it does not change the saturation level to above 45 mm and that it is not extremely heavy.

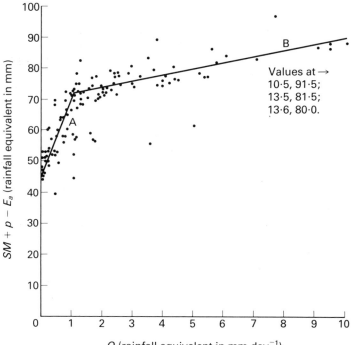

Values at →
10·5, 91·5;
13·5, 81·5;
13·6, 80·0.

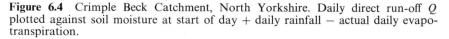

Figure 6.4   Crimple Beck Catchment, North Yorkshire. Daily direct run-off Q plotted against soil moisture at start of day + daily rainfall − actual daily evapotranspiration.

Since rainfall wets the vegetation canopy and the soil surface, it will return evapotranspiration to the full potential rate until the surface becomes dry again.

The moist state occurs with saturation levels between about 45 and 70 mm. It corresponds approximately to stage one of the soil-moisture model and evapotranspiration is at the full potential rate. Run-off is given approximately by the equation of curve A and will continue until the soil reaches the dry state. The boundary between the saturated and moist states represents the moisture level known as the field capacity of the soil. Soils are said to be at field capacity when excess water has drained away under gravitational action and the bulk of the water is held in the soil by capillary forces. The saturated state occurs with saturation levels above about 70 mm. The soil is now at field capacity and the vegetation may be wet. Additional rainfall runs off very quickly according to the equation of curve B.

Figure 6.3 illustrates the typical annual march of direct run-off in Britain. Direct run-off tends to be large in winter when the landscape is in either the moist or saturated condition and small in summer when it is mostly in the dry state. The main water losses are by run-off in winter and evapotranspiration in summer.

1.4 Rainfall

Rainfall is one of the fundamental inputs to climatological models so it is necessary to consider its distribution over the Pennines. Rainfall in the Pennines has two broad components, the first resulting from synoptic situations which are strongly influenced by relief while the second arises from synoptic situations which are not so influenced. Goh (1975) has found for the western Pennines that the greatest rainfall/elevation gradients are observed in cold fronts, and that in these situations the rainfall/elevation gradient increases rapidly with total rainfall. In contrast, over the eastern Pennines the relationship between rainfall/elevation gradients and storm type is less clear, with typical values between 0·01 and 0·04 mm increase in rainfall per storm per metre increase in elevation. This is probably because much of the orographically-influenced rainfall is formed on the western windward slopes and carried by the wind to the western side.

The relationship between relief and mean rainfall in the eastern Pennines has been investigated by Goh (1975) and Goh and Lockwood (1974). Spreen (1947) and others have shown that precipitation varies with such relief functions as elevation, slope and aspect. In the Pennines it was found that mean precipitation was mainly a function of mean relief and that factors such as aspect were relatively unimportant. For the period 1916–50, the mean rainfall in the eastern Pennines (y) is related to the mean elevation (x m) within an 8 km radius by:

$$y = 2 \cdot 023x + 564 \cdot 27$$

The equation for the western Pennines is:

$$y = 1 \cdot 890x + 916 \cdot 29$$

Seasonal values for the eastern Pennines are given by:

November to February	$y = 0.957x + 187.31$
March to April	$y = 0.265x + 73.98$
May to August	$y = 0.467x + 206.80$
September to October	$y = 0.425x + 93.82$

The exact seasonal or annual rainfall/relief gradient depends on the number and the nature of the storms. Normally, the more storms the greater the gradient, while fewer storms leads to a lower gradient and a generally more uniform annual rainfall distribution.

1.5 Results of a general climatic model
It is possible to express all the interactions and stores shown in Figure 6.1 in terms of equations and thus produce a general hydroclimatological model of the central Pennines. For given inputs of daily climatological data it is possible to produce daily evapotranspiration, run-off and soil moisture values and thus to estimate monthly and annual means.

Since plenty of observations of the duration of bright sunshine were available, radiation was expressed in terms of the percentage of normal sunshine. The relationship between global radiation and hours of bright sunshine was explained earlier. A storm sequence was generated which produced a good approximation to the average annual march of rainfall. Other climatological inputs were the appropriate mean values, usually for about 150 m.

The simplest results concern variable sunshine amounts with all other parameters, including rainfall, set at the climatological mean values. Variations in the number of sunshine hours per day implies variations in cloudiness and hence in global radiation. Some typical results are given in Table 6.2 which shows that as the number of sunshine hours increases

Table 6.2   Variation of direct run-off, groundwater re-charge and actual evaporation with sunshine. Rainfall constant at 795 mm per year; sunshine expressed as a percentage of the 1931–60 annual mean at Bradford; units: mm per year.

Percentage of normal sunshine	Direct run-off	Groundwater re-charge	Actual evapotranspiration
200	347	79	369
150	353	79	363
100	367	79	349
50	388	81	326
20	402	84	309

so also does the actual evapotranspiration, but with high sunshine hours the increase becomes small. This is because most evapotranspiration occurs in the summer (see Figure 6.2), and increases are eventually limited by a lack of soil moisture. Direct run-off slowly decreases as sunshine hours increase, but again at high sunshine values the rate of decrease is small. Most direct run-off occurs in winter when both sunshine and evapotranspiration totals are small, and so increases in summer sunshine when the soils are already dry for long periods can only have a small influence on annual run-off. Similar comments apply to the groundwater re-charge. The observed range of sunshine variation in the Pennines is in the region of 30 to 175 per cent of normal, so the values given in the table cover all possible variations.

Some results of varying rainfall, while other climatological parameters are kept constant, are shown in Figure 6.5. Three annual curves are presented; the lower one representing groundwater re-charge which on the annual scale is assumed to re-appear as river flow, the middle one direct run-off, and the upper one the total run-off which is the sum of the two lower curves. Also shown in Figure 6.5 are the average annual run-off values plotted against the 1916–50 average annual catchment rainfalls for various Pennine rivers. The correspondence between the measured run-off values and the predicted total run-off is reasonably good. Estimates of annual average base flows for four reasonably natural Pennine rivers are also shown in Figure 6.5, where there is a good correspondence with the groundwater re-charge/base flow curve.

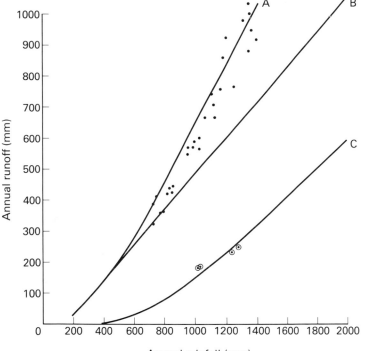

Figure 6.5 Predicted annual run-off (mm rainfall equivalent) for the central Pennines plotted against annual rainfall. Curve A, total predicted run-off; curve B, total predicted direct run-off; curve C, predicted groundwater re-charge/base flow. Dots show average actual discharge for east Pennine rivers. Circles show average actual base flows for natural Pennine rivers.

Figures 6.6 and 6.7 show the distribution of surface run-off and ground-water re-charge plotted against both precipitation and sunshine. These diagrams suggest that both surface run-off and groundwater re-charge are very closely controlled by precipitation, sunshine being of little importance.

The distribution of average annual rainfall with altitude in the Pennines is known and was described earlier. It is therefore possible to simulate the seasonal changes of both soil moisture and direct run-off with elevation assuming the month to month changes in rainfall observed in a year which is average for the period 1916–50. The results of such a simulation are shown in Figures 6.8 and 6.9. It is assumed that the sunshine was the average for Bradford (West Yorkshire), for it has already been demonstrated that small variations in sunshine are unimportant. Soil moisture values (Figure 6.8) are for the last day of the month, and therefore only indicate the broad pattern of the annual changes. A marked summer period with dry soils followed by a relatively rapid re-charge period in the autumn is evident for annual rainfalls below about 800 mm. At higher elevations and rainfalls, the summer period with dry soils is less marked and is broken by periods when the soil is rather wet. The direct run-off illustrated in Figure 6.9 shows a similar variation to the soil moisture. The very dry summer soils

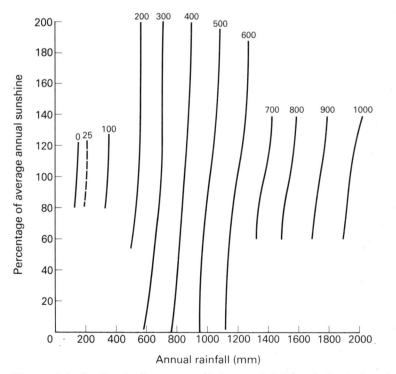

Figure 6.6 Predicted direct run-off (mm rainfall equivalent) for the central Pennines plotted against annual rainfall and sunshine. Sunshine in terms of Bradford average for 1931–60.

correspond to an extensive period of nil direct run-off, with run-off increasing as soil moisture increases in the autumn. Maximum direct run-off values are found in the winter.

2 The consequences of climatic changes

Kutzbach (1974) recently commented that a number of circumstances are combining to focus attention on the study of climate and climate variability. Global food production is at present not adequate to support properly the global population, and therefore the effect of weather and climate variability on crop production has acquired increased significance. According to Winstanley (1975), world food production must increase by 3·6 per cent per year to meet the demands imposed by increasing levels of affluence and by rapidly increasing populations. In actual fact the rate of growth of world food production slowed down from 3·1 per cent per year in 1952–62 to 2·7 per cent per year in 1962–72. The world's food supply is now perilously dependent on good weather and prevailing climatic conditions. Abrupt changes in the length of the growing season, annual rainfall, or mean temperature could disrupt modern and primitive agricultural systems alike.

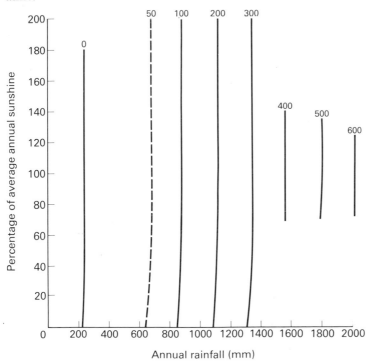

Figure 6.7 Predicted groundwater re-charge (mm rainfall equivalent) for the central Pennines plotted against annual rainfall and sunshine. Sunshine in terms of Bradford average for 1931–60.

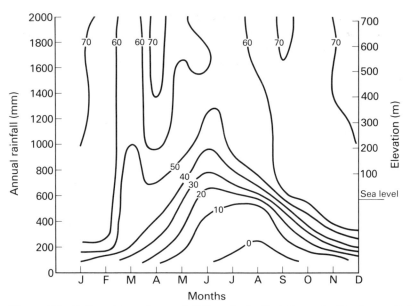

Figure 6.8   Soil moisture (mm rainfall equivalent) at the end of each month in the central Pennines assuming average march of rainfall. Elevation related to rainfall by an annual elevation/rainfall equation. Soil saturated at 70 mm.

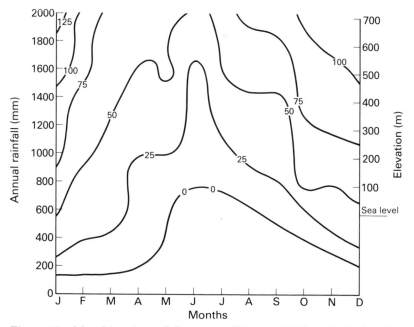

Figure 6.9   Monthly values of direct run-off (mm rainfall equivalent) in the central Pennines assuming average march of rainfall. Elevation related to rainfall by an annual elevation/rainfall equation.

Approximately 90 per cent of the world's food is grown and consumed locally (i.e. within national boundaries). About 10 per cent, however, is exported and imported, providing sustenance for the hungry half of the world, i.e. the 50 per cent or so of the population living in subequatorial and monsoon lands, the developing nations, and those where excessive populations out-strip local capacity to produce (Collis 1975). Cereals constitute the principal commodities traded, of which the United States contributes some 55 per cent, Canada 12 per cent, Australia 6 per cent and Argentina 1 or 2 per cent of the total shipments. Thus North America has been the only really substantial grain exporting region in the 1970's. Table 6.3 shows that this is a very recent state of affairs, and that in the 1930's most regions except western Europe were net exporters of cereals.

Table 6.3 The changing pattern of world grain trade (*based on US Department of Agriculture data compiled by Brown, 1975 and Schneider and Mesirow, 1977*). (Plus sign indicates net exports; minus sign, net imports.)

Region	1934–8	1948–52	1960	1970	1976*
		(million metric tons)			
North America	+5	+23	+39	+56	+94
Latin America	+9	+1	0	+4	−3
Western Europe	−24	−22	−25	−30	−17
Eastern Europe and USSR	+5	—	0	+1	−25
Africa	+1	0	−2	−5	−10
Asia	+2	−6	−17	−34	−47
Australia and New Zealand	+3	+3	+6	+12	+8

* Preliminary, fiscal year.

To assess the vulnerability of the world to fluctuations in food production, Brown (1975) compiled what he called an index of world food security for the years 1961–74, shown here in Table 6.4. The 'landbank' programme that kept US crop land idle in the 1960's and early 1970's was originally implemented to keep grain production from becoming too high and creating 'surpluses' that would depress domestic grain prices and reduce farmers' profit incentives.

According to Schneider and Mesirow (1976), consistently high crop yields depend on many factors: water, good seed stock, maintenance of genetic variability of crops, fertilizer, high productivity of the soils, pesticides, the pest-control services of natural ecosystems, good management skills of the farmer, capital to acquire and maintain technology, and stable climatic conditions. Improved technologies have led to steadily increasing crop yields during the present century and have tended to mask the influence of the weather.

McQuigg (1974) has provided some conclusive evidence on the effects of weather on agricultural output in the United States. In Figure 6.10, grain yields have been normalized to 1973 technology. Accordingly, variations in yields are solely a function of the variability of the weather from year to year.

Table 6.4   Index of world food security, 1961–76 (*prepared on the basis of US Department of Agriculture data, compiled by Brown, 1975 and Schneider and Mesirow, 1977*).

Year	Reserve stocks of grain	Grain equivalent of idled US cropland	Total reserves	Reserves as days of annual grain consumption
	(million metric tons)			
1961	163	68	231	105
1962	176	81	257	105
1963	149	70	219	95
1964	153	70	223	87
1965	147	71	218	91
1966	151	78	229	84
1967	115	51	166	59
1968	144	61	205	71
1969	159	73	232	85
1970	188	71	259	89
1971	168	41	209	71
1972	130	78	208	69
1973	148	24	172	55
1974	108	0	108	33
1975	111	0	111	35
1976*	100	0	100	31

* Preliminary.

In this figure, 'normal' weather is defined as long-term average precipitation and temperature conditions. Drought as indicated here represents conditions resulting in 10 per cent lower yields than 'normal' weather. 'Normal' weather represents generally optimum growth conditions, drier and wetter, hotter or colder conditions tending to produce reduced yields. The period of severe drought in the central United States during the 1930's, which was characterized by above-normal temperatures and below-normal precipitation, shows up in Figure 6.10 as several years of low grain yields. The really interesting fact shown by Figure 6.10 is that the weather of the period from approximately 1955 to 1970 was especially favourable to grain production in the United States.

Collis (1975) suggests that the favourable weather of the 1955–70 period (which to a greater or lesser degree was widely experienced), coming as it did at a period of great technological progress, has had a tremendous, but unrealized influence on the thinking of US agricultural economists and national policy makers. On the basis of such favourable experience the possibility of adverse weather has been underestimated. On a world-wide basis after the poor weather of 1972, there were good conditions in 1973 which were, however, unable to restore stability in the face of continued rises in demand. It was followed by the serious shortfall in 1974 production in many areas due to adverse weather. In contrast, US grain production in 1976 and 1977 was good, with a return to large surpluses.

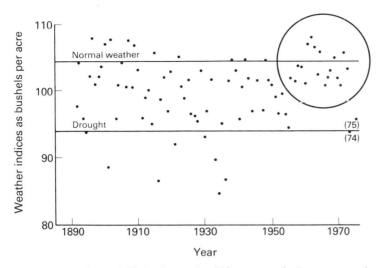

Figure 6.10 Corn yields in five major US corn-producing states under the assumptions that technology is fixed at 1973 levels and the only influence on corn yield variability is the weather patterns that prevailed from 1890–1975. This technique isolates the impact of weather on year-to-year crop variability, and demonstrates that the weather in the corn belt from 1956–73 was abnormally good (*after McQuigg, 1974, Collis, 1975, and US National Academy of Sciences, 1976*).

2.1 Drought

One of man's worst natural enemies is drought. Its beginning is subtle, its progress is insidious and its effects can be devastating (Hounam *et al.* 1975). Since rainfall is still the main source of fresh water for agricultural, domestic and industrial use, drought can have an impact ranging from slight personal inconvenience to endangered nationhood.

A study of drought requires an objective definition, but according to Hounam *et al.* no universally acceptable definition has so far been developed. It can be assumed that the basic cause of drought is inadequate precipitation, but the exact meaning of the word 'inadequate' needs careful definition. The distribution of average annual precipitation does not by itself give an indication of drought incidence or intensity although obviously it has a marked control over normal land use. The use of such a distribution may lead to areas of low average rainfall being identified as drought areas. Aridity is usually defined in terms of low average rainfall and, ignoring the possibility of climatic change, is a permanent climatic feature of a region. Hounam *et al.* suggest that drought is a temporary feature in the sense that, considered in the context of variability, it is experienced only when rainfall deviates appreciably below normal. Aridity is, by definition, restricted to regions of low rainfall, and usually of high temperature, whereas drought is possible in virtually any rainfall or temperature regime. Activities in arid zones are geared to meet the 'permanence' of aridity but a drought situation results in at least some interruption of normal activities in all zones.

Drought should be regarded as a hydrologically extreme state in exactly the same manner that flooding is regarded as a state of extreme river flow. In many ways drought is the hydrological opposite to flooding. Dry spells with large soil moisture deficits are normal at most lowland middle-latitude sites in late spring and summer. Such dry spells should not be referred to as droughts unless they reach an unusual intensity or are abnormally long for the particular sites. The analogy with river flow is close, since a river flood is only said to occur when the rate of run-off is unusually high, and this particular state will vary from river to river. Thus drought should be defined in terms of prolonged and abnormal moisture deficiency.

One major reason for the lack of a universally agreed definition of drought is that concepts of drought are related to particular water uses. For instance, British agriculture is chiefly concerned with adequate summer rainfall to offset the evaporation and transpiration which occurs in that season and gives rise to large soil moisture deficits. On the other hand, British water supply interests lie much more with winter rain which provides run-off for reservoirs and percolation to re-charge aquifers. It is seen in Figures 6.6, 6.7 and 6.9 that both run-off and groundwater re-charge are almost completely a function of the winter rainfall. On a world scale, ideas on agricultural drought vary depending on the exact type of agricultural activity and the normal local climate.

Agricultural drought may be expressed in terms of the degree to which growing plants have been adversely affected by an abnormal soil moisture deficiency. The deficiency may result either from an unusually small moisture supply or an unusually large moisture demand. While a deficiency in the water supply for livestock may be regarded as a facet of agricultural drought, it is really a different problem because it does not depend primarily on soil moisture.

Since every crop has its own drought-sensitive periods, a proper analysis of agricultural drought should cover each crop separately. Tabony (1977) has considered grassland drought in Britain, where grass covers about 70 per cent of the agricultural land. Grass growth becomes restricted when the availability of water from the upper layers of the soil becomes restricted as described earlier in this chapter. Tabony therefore suggests that a possible measure of 'grassland' drought D_a is given by:

$$D_a = E_p - E_g$$

where E_p is the potential evapotranspiration and E_g is the transpiration effective for growth. Tabony obtained E_g from E_p by using 25·4 mm (1 inch) root constant.

In contrast, hydrologists in Britain are interested in adequate winter rains to fill reservoirs and re-charge aquifers. The hydrologically effective rainfall (R_e) for such purposes is the water surplus remaining after evaporation has taken place and any soil moisture deficit has been removed. This is seen clearly in Figure 6.2. Therefore, Tabony considers that R_e is a possible measure of 'hydrological' drought.

Wigley and Atkinson (1977) have used values of soil moisture deficit to define agricultural drought. Long homogeneous series of precipitation and evapotranspiration data for Kew (51°28′ N, 0°19′ W) going back to 1698 have been constructed and discussed by Wales-Smith (1971, 1973a, 1973b);

Plate 11 The reservoir at Pitsford, Northampton, 5 May 1976. Taken during the worst British drought for 250 years, the photograph shows the dried up bed of the reservoir with water levels 7 m below normal. *Reproduced by permission of Keystone Press Agency Ltd.*

these are the longest such series available anywhere in the world. Wigley and Atkinson have constructed soil moisture deficit values back to 1698 using the Kew data. They averaged the soil moisture deficit over the whole growing season and used a composite of short- and long-rooted vegetation results in order to provide a general agricultural drought index. Using this index, the 14 most severe agricultural droughts over the interval 1698–1976 are: 1705, 1731, 1762, 1781, 1844, 1893, 1901, 1921, 1934, 1938, 1944, 1965, 1974, 1976. The seven worst years are, in order, 1976, 1934, 1944, 1893, 1938, 1921, 1974. 1976 shows up as the worst agricultural drought since at least 1698, although only fractionally more severe than 1934. However it is probably more realistic to state that 1976 was among the three worst agricultural droughts during the 279 year period of record. Ten-year running means of soil moisture deficit averaged over the growing season are shown in Figure 6.11.

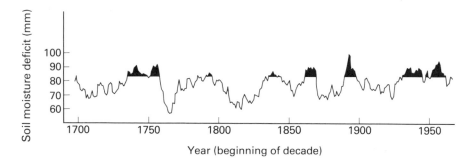

Figure 6.11   Ten-year running means of soil moisture deficit averaged over the growing season at Kew. The value of year *n* corresponds to the decade *n* to *n* − 9 inclusive. An arbitrary datum level of 84 mm has been shown to accentuate the periods of higher deficits (*after Wigley and Atkinson, 1977*).

Tabony (1977) has listed the ten most severe grassland drought at Kew from 1871 to 1975, obtained from his index D_a. These are in order of magnitude: 1959, 1921, 1893, 1972, 1975, 1949, 1911, 1938, 1899, 1933. He comments that calculations of D_a had not been made at the time of publishing his paper, but that the summer of 1976 is likely to compare with those of 1939 and 1921. The fact that his list does not agree with that of Wigley and Atkinson is an indication of the extreme difficulty in designing drought indices, and shows that the whole problem must be approached with some caution. It is possible that both authors are identifying groups of dry years rather than specific individual years. The ending of the main meteorological drought of the summer of 1976 occurred in the last few days of August or in early September in most parts of the United Kingdom. Murray (1977) has listed in Table 6.5 the rainfall (as percentages of average) for the United Kingdom for various periods from 3 to 18 months ending in August 1976. Over England and Wales combined the rainfall was slightly less in the 3 month period June to August in 1800 than in 1976, and in the 6 month period March to August in 1741 than in 1976, but for the other periods shown in Table 6.5 the rainfall over England and Wales was the lowest on record for the 9, 12, 15, 16 and 18 month periods with return periods of at least 250 years.

Table 6.5 Rainfall, as percentage of average (1916–50), for various periods from 3 to 18 months ending in August 1976 over parts of and the whole of the United Kingdom (*after Murray, 1977*).

	3 months June 76– Aug. 76	6 months Mar. 76– Aug. 76	9 months Dec. 75– Aug. 76	12 months Sept. 75– Aug. 76	15 months June 75– Aug. 76	16 months May 75– Aug. 76	18 months Mar. 75– Aug. 76
United Kingdom	41	64	68	74	73	73	76
England and Wales	35	52	55	63	63	64	70
England	37	52	55	63	63	64	70
Wales	27	50	57	64	64	64	68
Scotland	48	79	83	86	85	83	84
Northern Ireland	48	74	74	80	76	73	75

Plate 12 Sahel drought in Niger, July 1973. Feeding the family's few remaining goats at an encampment near a desert village in Central Niger. *Reproduced by permission of FAO, Rome.*

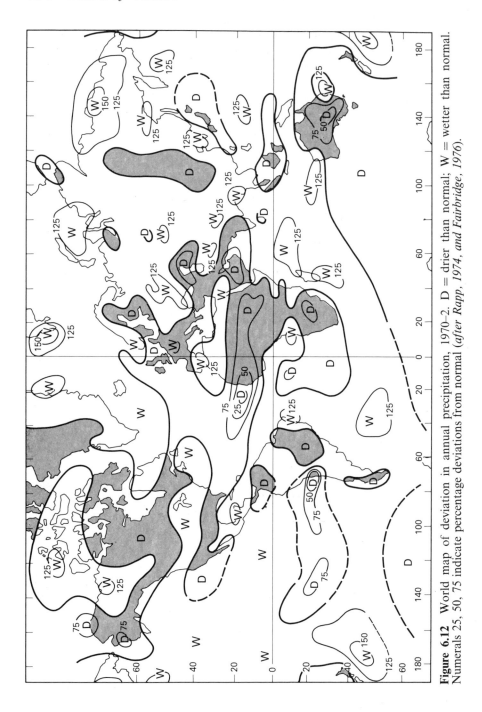

Figure 6.12 World map of deviation in annual precipitation, 1970–2. D = drier than normal; W = wetter than normal. Numerals 25, 50, 75 indicate percentage deviations from normal (*after Rapp, 1974, and Fairbridge, 1976*).

2.1.1 *The drought in the West African Sahel*

Figure 6.12 shows the deviation in annual precipitation for the period 1970–2. It represents a short-term climatic fluctuation, with trends towards desiccation in tropical Africa, western India, most of Australia, western Europe and the southern United States. The dry years of the 1970's over Britain are clearly shown. Among the worst droughts is that observed over the West Africa Sahel region, and this is now discussed in detail.

The main characteristics of the climate of the West Africa Sahel (see Figure 6.13) is the alternation between wet and dry seasons, since in the zone between 10° and 20° N rain falls only from May or June until September or October. The isohyets are aligned east–west and annual rainfall decreases rapidly from about 1000 mm at 10° N to 50 mm or less at 20° N. The length of the growing season correspondingly decreases from about five months at 10° N to two months at 17° or 18° N. Winstanley (1975) remarks that almost 90 per cent of the active labour force in Mauritania, Senegal, Mali, Upper Volta, Niger and Chad is engaged on agriculture, livestock rearing and fishing, mainly at a subsistence level.

Five-year means of annual rainfall in the Sahel region of West Africa expressed as percentages of the 1905–70 average are shown in Figure 6.14.

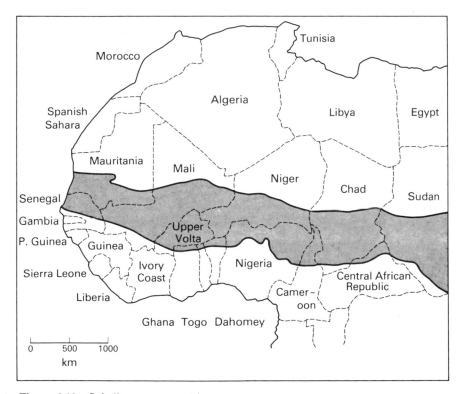

Figure 6.13   Sahelian zone countries.

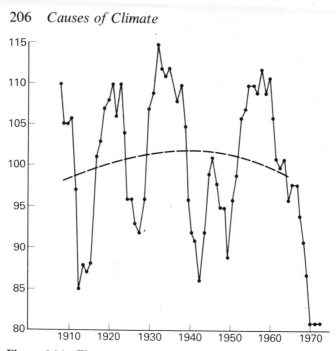

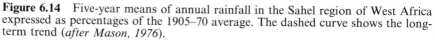

Figure 6.14   Five-year means of annual rainfall in the Sahel region of West Africa expressed as percentages of the 1905–70 average. The dashed curve shows the long-term trend (*after Mason, 1976*).

Series of dry years are recorded for 1900–3, 1909–16 (especially 1913–14), 1931, 1940–4, and then 1968–73. These do not correspond with the series of dry years on the subtropical, Mediterranean edge of the Sahara which are for the most part the reverse. They have not been of the same severity in the whole of the Sahel, except for the most recent drought. Rainfall for the period 1931–60 is taken almost universally as the standard for comparison for the Sahel region. This is unfortunate, since Figure 6.14 shows that it was a period of greater than 'average' rainfall in West Africa.

. During the period 1968–72 rainfall along the desert fringe was only 40 to 60 per cent of the 1931–60 average, with the largest percentage decreases in monthly rainfall occurring at the beginning and end of the wet season. Winstanley (1975) gives a number of other interesting facts about the latest Sahelian drought. From 1967 to 1972–3 the discharge of the Niger and Senegal Rivers declined by 50 and 70 per cent respectively, and Lake Chad was reduced in area by 65 per cent. The level of the water table also fell and the annual floods on the inland deltas of the Niger and Senegal Rivers virtually disappeared. In 1974, rainfall became sufficient again and the pastures revived quickly.

If a catastrophe is a situation in which more than 100 people die and/or 1,000 are injured and/or $1,000,000 worth of damage is sustained, then this disaster in the Sahel is a parallel to recent disasters in Bangladesh in terms of magnitude of death, disability and destruction (UN protein advisory group 1974). Six years of drought have brought the sub-Saharan areas of

Plate 13   Sahel drought in Upper Volta, June 1973. A hungry goat stripping what little green is left on thorn trees in northern Upper Volta. *Reproduced by permission of FAO, Rome.*

the Sahel, parts of Ethiopia, and Somalia to nothing short of catastrophe. More than 25 million people have been affected and initial estimates for West Africa suggest that more than 100,000 people have died (Sheets and Morris 1974). Although these figures have been queried this seems the most realistic estimate (O'Keefe and Wisner 1975). Cattle losses are placed around 75 per cent in Mauretania, 50 per cent in Senegal, 50–80 per cent in Mali, 50–100 per cent in Upper Volta, 80 per cent in Niger and 90 per cent in Chad. Production of millet, sorghum, corn, rice and groundnuts fell by 40 per cent and the total *per capita* food production by 35 per cent.

The Sahel, a term derived from the Arabic word for border, was once one of the more important areas of Africa and the home of several trading empires. The key to the Sahelian way of life was a remarkably efficient adaptation to the semi-desert environment which has been described by Wade (1974). The dry season finds the nomads with their cattle herds as far south as they can go without venturing within the range of the tsetse fly. With the first rains, the grass springs up and the herds move northwards. The migration continues as long as the grass ahead looks greener than that at hand, until the northern edge of the Sahelian rain belt is reached. When the grass is eaten off, the return to the south begins. This time the cattle are grazing a crop of grass that grew up behind them on their way north, and they are drinking standing water remaining from the rainy season. Back in their dry-season range the cattle find a crop of mature grass that will carry them for 8 or 9 months to the next growing season.

According to Wade, western intervention has made itself felt in many ways, some inadvertent, some deliberate. The French colonial division of the Sahel into separate states has faced the nomad tribes with national governments which have tried to settle them, tax them and reduce their freedom of movement by preventing passage across state boundaries. The impact of Western medicine has led to a general increase in both human and animal populations. The people of the Sahel are increasing at a rate of 2·5 per cent a year, one of the highest rates of population increase in the world. Also, according to the FAO, the number of cattle grew from about 18 to 25 million between 1960 and 1971. The optimum number, according to the World Bank, is 15 million. Thus too many cattle with no place to go has led to the overgrazing and destruction of the Sahelian pasturelands.

The Ekrafane ranch is quoted by Wade as an example of the phenomenon of desertification. This was first noticed as a curiously-shaped green pentagon on a NASA satellite photograph. A visit to the site of the pentagon showed that the difference between it and the surrounding desert was nothing more than a barbed-wire fence. Within was a 250,000 acre ranch, divided into five sectors with the cattle allowed to graze one sector a year. Although the ranch was started at the same time as the drought began, the simple protection afforded the land was enough to make the difference between pasture and desert.

While the herders were overtaxing the pastures, the farmers were doing the same to the arable land. Population increase led to more and more people trying to farm the land, while the French introduced cash crops to earn foreign exchange. In many cases the ecologically fragile lands could not take the strain of intensive agriculture, the soil started to lose its structure, and desertification began.

It is difficult to speculate on the future of the West African Sahel. The ideas of Charney *et al.* (1976) on the growth of deserts due to overgrazing were discussed in Chapter 3. Theories of this type suggest that the drought may have been intensified by the destruction of the vegetation by overgrazing and poor land cultivation. Thus a downward spiral may have been created with decreasing rainfall leading to a change to desert conditions which in turn help to further depress the rainfall. The first years of independence coincided with a sequence of wet years up to 1968. The future rainfall will probably be higher than in the period 1968–73, but still not as high as in the early years of independence.

3 The climatic future

In recent testimony before the Subcommittee on Environment and the Atmosphere of the Committee on Science and Technology, US House of Representatives, Knox and MacCracken (1976) commented that 'there is widespread agreement that mankind is on the verge, if not already capable of being able to take actions (e.g., emit pollutants in sufficient quantity) that may noticeably disturb the delicate atmospheric balances of energy, radiation and water on the regional and, perhaps, global scale'. Table 6.6 lists some of the pollutants and briefly describes their importance. Essentially, all act to disturb either the radiation balance or the precipitation mechanism, two processes having, as already described, fundamental importance in determining the climate.

According to Knox and MacCracken, a number of other possible mechanisms for modifying the climate that do not involve pollutants but that do result from the activities of mankind are:

(a) diverting rivers flowing into the Arctic Ocean which may cause a melting of Arctic Ocean sea-ice;
(b) irrigating arid lands;
(c) changing land surface reflectivity (albedo) by urbanization or agricultural practices;
(d) clearing tropical rain forest;
(e) overgrazing of arid lands.

On a global scale, there is no convincing evidence that any action of mankind has yet affected the global climate. Thus, estimates of potential effects of future emissions must be based on theoretical analysis, often using computer-run numerical models. Further, Landsberg (1976) states that cause and effect relationships of climate are known in only a rudimentary way; hence predictions for either short or long intervals are afflicted with great uncertainties.

Table 6.6 Pollutants and their potential atmospheric effects (*after Knox and MacCracken, 1976*).

Pollutant and source	Observed trend	Potential atmospheric effect	Status of assessment capability	Time scale of importance
Carbon dioxide (CO_2) from combustion of fossil fuels.	Up more than 20% in last 100 years.	Increased global temperatures leading to melting of polar ice-caps, sea-level increase, and perturbations of marine biology.	Numerical model assignments of the global average effect on temperature differ by about a factor of 2; consequence chains need more study.	Thorough assessment needed in next five years; may be a problem over next 50 years.
Fluorocarbons (e.g., freon) from aerosol cans, refrigeration systems, etc. Nitrogen oxides from high-flying aircraft (and perhaps from fertilizers).	Fluorocarbons are now detectable throughout the atmosphere. Nitrogen oxides are a natural component. Stratospheric measurement programme is being established to determine levels and trends.	Reduction of the global stratospheric ozone layer and perturbation of the atmosphere's radiation balance. Analysis of current trend in ozone is not yet definitive due to natural variability.	Numerical models are capable of assessing the order of magnitude of the various effects, with uncertainties related to lack of basic information on reactants, reactions and reaction rates; the natural chlorine and nitrogen balance; and the limitations in simulating simultaneously global chemistry, transport, and seasonal and diurnal processes.	Initial assessment in progress by National Academy of Sciences; action probably needed within several years.
Krypton-85 from nuclear	Building up	Modification of the	Not adequate	Thorough assessment

Source	Present concentration/trend	Possible effects	Current knowledge	Evaluation needed
...power plants.	...proportionally with nuclear power generation.	...field, which may cause modification of the hydrologic cycle.		needed, may be a problem over next 100 years with growth of nuclear power industry.
Sulphur compounds from fossil fuel combustion.	Not well established, but concentrations may already be too high on occasion.	May affect regional precipitation chemistry and acidity on regional to subcontinental scale.	Sulphur balance not well understand.	May presently be a problem that would be aggravated by further coal burning.
Dust from combustion, slash/burn agriculture, and improper land conservation.	Not well established because of evolution of sources and particle sizes with controls.	Initial response is temperature change (sign dependent on location and source type) and precipitation modification. Problem mainly on subcontinental but possibly up to global scale.	Theoretical capability is improvement, but inadequately knowledge of both trends and consequences exist.	Further evaluation needed as improved data become available.
Heat and water releases to the atmosphere from the energy generation process (thermal pollution, cooling towers, etc).	Increasing with energy generation.	Temperature and precipitation modification on the local and regional scale.	Models of atmospheric boundary layer are being developed.	Evaluation needed in regions of concentrated energy generation (e.g., energy parks, etc).
Oceanic oil slicks from tanker cleaning, etc.	Not known.	By changing the reflectivity and evaporation characteristics of large oceanic areas, the earth's energy balance might be perturbed in an unknown way.	Further research needed.	Assessment needed as capability for evaluation improves.

Plate 14 Wind blown dust is often grouped among the natural pollutant sources, but in many instances it has been at least enhanced by anthropogenic influences. American Secretary of Agriculture Ezra Taft Benson visiting the Dakota Dust Bowl in April 1955, where dry soil from ploughed fields is being blown into blinding dust storms. *Reproduced by permission of Camera Press Ltd.*

3.1 Global pollution

3.1.1 *Global pollutant emissions in the troposphere*
In the case of particulate pollutants it is customary to distinguish between man-made (anthropogenic) and natural particle emissions. Atmospheric particulates may originate from either the direct emission of particulate material into the atmosphere from anthropogenic (e.g. urban/industrial processes) and natural sources (e.g. volcanism) or from particle formation by gas reactions. The main sources of direct particle production include sea-salt spray, wind blown dust, volcanic emissions, meteoric debris and forest fires. Also large amounts of water vapour, carbon dioxide, sulphur dioxide, carbon monoxide, hydrogen and halides, especially hydrogen chloride, are emitted to the atmosphere and some of these are converted into various types of solid particles and liquid drops.

According to Bach (1976) the bulk of the directly-formed particle material in the atmosphere stems from eolian forces which transport particles that originate on the surface of the earth either over land (wind blown dust) or from sea to land (sea-salt spray) and from volcanic forces, which eject particles from inside the earth into the atmosphere. The direct particle contribution from outer space in the form of meteoric debris is rather small. Forest fires also produce sizeable amounts of particles, but it is difficult to decide how much is produced naturally and how much under the influence of man. Also wind blown dust has been grouped here among natural pollutant sources, but in many instances it has been at least enhanced by anthropogenic influences. For example, the dust storms on the Great Plains of the United States are to some extent related to agricultural activities.

Plate 15 Anthropogenic air pollution from steel mills near Cleveland, Ohio. *Reproduced by permission of Grant Heilman, Lititz, P.A.*

Sulphate aerosols can be generated by the oxidation of various gaseous emissions. Most of the natural sulphate in the air seems to be produced from decaying vegetation and animals either on land or in the oceans. Similar aerosols are generated from hydrocarbons and nitrous oxides emitted by vegetation.

Estimates of global particle production from natural and man-made sources are shown in Table 6.7. It is seen that the direct anthropogenic annual global particle production amounts to only 6·8 per cent of the direct particle emission from natural sources. Also, only 27 per cent of the total man-made particulates are of primary origin, since more than 70 per cent are in the form of gaseous precursors. The anthropogenic contribution to the total global particulate production is 15·2 per cent.

Table 6.7 Estimates of global particle production from natural and man-made sources (*after Bach, 1976*).

Source	After Peterson and Junge (1971)		After Hidy and Brock (1971)	After SMIC report (1971)*
	All sizes	<5μm		
	Man-made			
Direct particle production				
Transportation	2·2	1·8		
Stationary fuel sources	43·4	9·6		
Industrial processes	56·4	12·4		
Solid waste disposal	2·4	0·4		
Miscellaneous	28·8	5·4		
Subtotal	133·2	29·6	36·8–110	10–90
Particles formed from gases				
Converted sulphates	220	200	109·5	130–200
Converted nitrates	40	35	23	30–35
Converted hydrocarbons	15	15	27	15–90
Subtotal	275	250	159·5	175–325
Total man-made	408	280	269	185–415
	Natural			
Direct particle production				
Sea salt	1000	500	1095	300
Wind blown dust	500	250	7–365	100–500
Volcanic emissions		25	4	25–150
Meteoric debris	10	0	0·02–0·2	
Forest fires	35	5	146	3–150
Subtotal	1545	780	1610	428–1100
Particles formed from gases				
Converted sulphates	420	335	36·8–365	130–200
Converted nitrates	75	60	600–620	140–700
Converted hydrocarbons	75	75	182–1095	75–200
Subtotal	570	470	2080	345–1100
Total natural	2115	1250	3690	773–2200
Grand total	2523	1530	3959	958–2615

The units are 10^6 tons per year.

* Values are for particles <20 μm.

The notion that more than 90 per cent of the major anthropogenic pollutants originate from industrial operations in the northern hemisphere needs some revision according to Bach (1976). Observations seem to indicate that tropical grassland fires are much greater contributors to global emission totals than had been hitherto believed. For instance, Table 6.8 shows that the United States contributes about 7 per cent to the global man-made particle pollution, while tropical agricultural burning contributes about 24 per cent to the total anthropogenic particle emission.

Table 6.8 Summary of United States, tropical and global emission estimates (*after Bach, 1976*).

			Total emission estimates (10^6 tons per year)		
Constituents	United States, 1972	Tropical agri-cultural burning	Global man-made	Global natural	Global total
CO	97	290	190–640	68–5,000	258–5,640
CO_2		7222	11,793–16,329	127,005–907,180	138,798–923,509
HC	25	72	80–90	90–435	170–525
CH_4			14–210	330–2,086	344–2,296
SO_2	30		62–133	40	62–173
H_2S			3	100	103
NO				390	390
NO_2			48	453–696	453–744
NO_x	22	7	45	453	498
NH_3			4	1,052–5,352	1,052–5,356
N_2O				144–589	144–589
Particulates	18	62	185–415	773–3,690	958–4,105

Mean residence times of aerosol particles have been compiled by Flohn (1973) for various layers in the atmosphere, and are shown in Figure 6.15. In agreement with SCEP (1970) this figure shows residence times of 1–4 days near the surface, 5–14 days in the lower and upper troposphere, 1–2 years below 20 km, and 3–10 years in the upper stratosphere and mesosphere. For a removal rate linearly proportional to the concentration the overall residence time is ordinarily given as the time required for the material to be reduced to $1/e$, or 37 per cent, of its original concentration. This means that much of the material may still remain in the atmosphere beyond the so-called residence time. The smallest particles are quickly removed by diffusional deposition and rainfall, whilst the largest particles are rapidly deposited by sedimentation. Particles in the intermediate size range ($0·1–1·0$ μm) having only a small sedimentation rate, a small inertial, and a small diffusional removal rate, remain much longer in the atmosphere than either the smallest or the largest aerosols. Condensation of vapour and the deposition of gases such as nitrogen oxides, sulphur compounds and hydrocarbons appear to be important factors not only in the growth and removal of particles but also in the removal of the gases themselves.

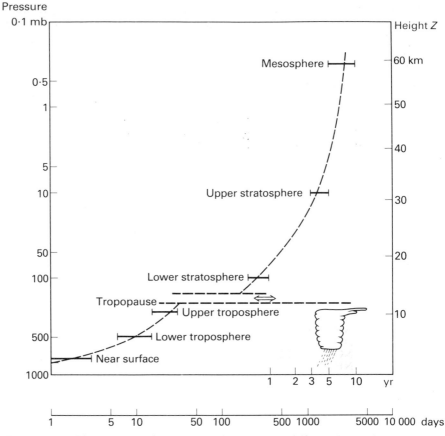

Figure 6.15 The mean residence times of aerosols in different layers (*after Flohn, 1973*).

3.1.2 *Global pollutant concentrations in the troposphere*

In Table 6.9, Ludwig *et al.* (1970, 1971) have tabulated for the United States the concentrations of suspended particulates and their major constituents for 1966–7. Thus the concentrations between urban stations, non-urban intermediate stations (reflecting agricultural activity), and remote stations (isolated from man's activities) can be compared. The comparison indicates that remote locations have background values for suspended particulates, soluble organics, and the converted aerosols that are typically about one fifth of the urban concentrations, Bach and Daniels (1975) have evaluated in Figure 6.16 the urban and non-urban concentration trends for suspended particulates for a number of stations in the United States for the period 1960–71. The bold and light vertical lines in Figure 6.16 denote the urban and non-urban concentration ranges. The bold solid and light solid horizontal curves represent the annual averages for the indicated number of urban and non-urban stations. There is a downward trend in the annual

Table 6.9 Selected particulate constituents as percentages of gross suspended particulates (1966–7), United States (*data from Ludwig et al., 1970*). (The number of stations is shown in brackets.)

	Urban				Non-urban			
	217 stations		Approximate (5)		Intermediate (15)		Remote (10)	
	μg m^{-3}	%	μg m^{-3}	%	μg m^{-3}	%	μg m^{-3}	%
Suspended particulates	102·0		45·0		40·0		21·0	
Benzene soluble organics	6·7	6·6	2·5	5·6	2·2	5·4	1·1	5·1
Ammonium ion	0·9	0·9	1·22	2·7	0·28	0·7	0·15	0·7
Nitrate ion	2·4	2·4	1·40	3·1	0·85	2·1	0·46	2·2
Sulphate ion	10·1	9·9	10·0	22·2	5·29	13·1	2·51	11·8
Copper	0·16	0·15	0·16	0·36	0·078	0·19	0·060	0·28
Iron	1·43	1·38	0·56	1·24	0·27	0·67	0·15	0·71
Manganese	0·073	0·07	0·026	0·06	0·012	0·03	0·005	0·02
Nickel	0·017	0·02	0·008	0·02	0·004	0·01	0·002	0·01
Lead	1·11	1·07	0·21	0·47	0·096	0·24	0·022	0·10

geometric mean values at urban stations, but non-urban stations show no discernible trend.

Long-term trends of atmospheric particle loading can be inferred from conductivity and solar radiation measurements of atmospheric turbidity. Using measurements of solar transmittance at Mauna Loa, Hawaii, Ellis and Pueschel (1971) have found no evidence of human activities. Figure 6.17 shows in fact no systematic trend of solar transmittance for the eruption-free period 1958–63. With the onset of the Mount Agung eruption in 1963, the solar transmittance decreased by about 1–5 per cent only to recover to pre-Mount Agung levels by the end of 1970. Bach (1976) comments that on the basis of stratospheric residence times of 1–2 years, the recovery to normal levels should have been sooner. The reason for this delay was probably the enhanced volcanic activity after the Agung eruption. From turbidity measurements taken in the Antarctic, Fischer (1967) has also concluded that no pronounced changes in atmospheric dust loading occurred from 1949 to 1966. In contrast, Suraqui *et al.* (1974) report a loss of nearly 10 per cent in the direct-beam solar radiation at Mt St Katherine, Sinai, over a period of 40 years. Between 1933 and 1937, Abbot and his co-workers from the Smithsonian Institution made normal-incidence measurements of solar radiation from Mt St Katherine at 2643 m above sea-level. The site was chosen by the Smithsonian because of its exceptionally clear skies. Measurements made 40 years later with newer instruments and one of the original silver disks used by Abbot indicate a loss of some 10 per cent of the solar energy contained in the direct beam. Since the nearest pollution source is the oil fields at Abu Rodeis some 40 km to the west, this indicates a relatively drastic change in solar insolation.

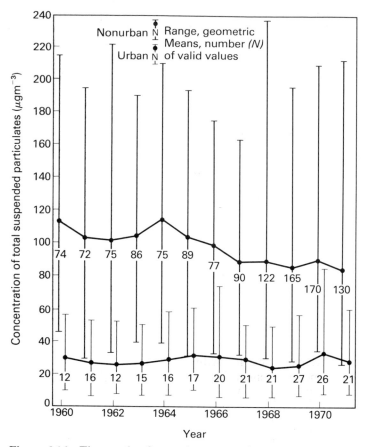

Figure 6.16   The trends of annual means and ranges of suspended particulates at urban and non-urban NASN stations in the United States, 1960–71 (*after Bach and Daniels, 1975*).

Observations of atmospheric conductivity over the ocean have been made from research vessels since the turn of the century, and summarized by Cobb and Wells (1970), and Cobb (1973). These show that there have been decreases in conductivity in this century (corresponding to increases in aerosol content near the surface) in the western North Atlantic, the western North Pacific, and the Indian Ocean close to India. For the North Atlantic over the period 1907 to 1969, Cobb and Wells found a 20 per cent reduction in conductivity, which is equivalent to a doubling of the fine particle content of the air.

The evidence for trends in atmospheric aerosol content is therefore contradictory, and this probably arises from differing local policies in pollution control. Kellogg *et al.* (1975) comment that aerosol trends have been generally upward in most of Europe and the Soviet Union, where increased industrialization has taken place and where pollution abatement efforts

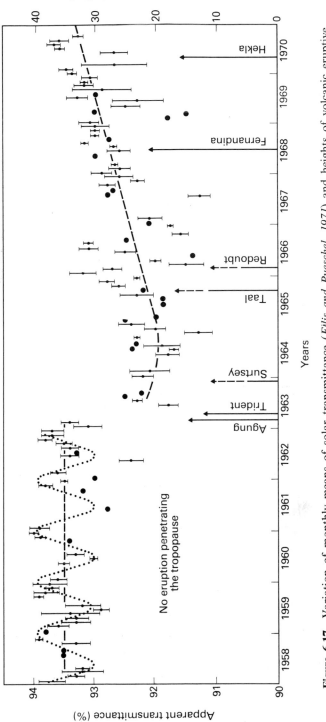

Figure 6.17 Variation of monthly means of solar transmittance (*Ellis and Pueschel, 1971*) and heights of volcanic eruptive clouds. (*Cronin, 1971*).

(before the 1970's) have apparently lagged. In the United States, Britain and some other countries, where vigorous pollution abatement efforts were already well under way in the early 1960's, the reduction seems to have kept up approximately with the increased industrial activity in the past decade or more.

The observations described above are so sparse that it is impossible to draw a global pattern of mean man-made aerosol pollution from them alone. Therefore, Kellogg *et al.* (1975) have attempted a very crude estimate of these patterns based on the considerations listed below.

(1) Aerosol production in a country is proportional to its general industrial activity, and furthermore that this activity can be measured by its gross national product (GNP). That is, gross national pollution is assumed to be proportional to GNP.

(2) Man-made aerosols, with a few trivial exceptions, are released at or within a few hundred metres of the surface, and will therefore be carried more or less with the prevailing surface winds.

(3) Aerosols at mid-latitudes have an average residence time of about a week at cloud level and 3 to 4 days near the surface. An average residence time for man-made aerosols was assumed to be 5 days. Taking the wind speed to be 5 m s^{-1}, the average travel of a particle will be about 2,000 km;

Plate 16   Anthropogenic air pollution from an oil refinery near Billings, Montana. *Reproduced by permission of Grant Heilman, Lititz, PA.*

stated another way, the aerosol plume from an industrial region will be depleted to $1/e$ (37 per cent) of its initial density after drifting 2,000 km.

Kellogg *et al.* (1975) have constructed maps on these assumptions, and they are reproduced in Figure 6.18. It is seen that aerosols generally lie in a belt at mid-latitudes in the northern hemisphere, the major exception being the North Pacific. There is little cross-equatorial flow, and the countries of the southern hemisphere contribute far less than the industrialized countries of the northern hemisphere. United States aerosols are almost entirely depleted before they reach Europe, but on the other hand, European and Soviet aerosols blanket most of Africa north of the equator in January and spill out over the tropical Atlantic. The reason for the lack of any detectable trend in tropospheric aerosol concentrations at Mauna Loa, Hawaii, while increases are reported from elsewhere, now becomes apparent, since very little man-made aerosol is finding its way into the North Pacific.

Of the gases perhaps SO_2 and H_2S, and CO_2 are the most important ones regarding climatic change, because the former act as precursors for aerosols with a net warming or cooling effect, while the latter always has a warming effect. The most important of these gases is CO_2, and Figure 6.19 shows a comparison of trends in atmospheric CO_2 for Mauna Loa, Hawaii, and the South Pole. Both the Antarctic and Mauna Loa show an annual increase in CO_2 concentration of the order of 1 ppm.

The production of CO_2 from such natural sources as the decay of organic matter and respiration of living organisms is balanced on a net annual basis by photosynthetic consumption. It is generally agreed that it is largely the burning of fossil fuels that is responsible for the continuous increase of CO_2 in the atmosphere. Bolin (1977) has claimed that part of the increase is also due to the expansion of forestry and agriculture. The increase in CO_2 concentration observed at the South Pole is approximately one half of what would have occurred if all of the CO_2 released by the burning of fossil fuel and limestone (to make cement) had remained airborne. It is interesting to note that the South Pole CO_2 concentrations are in general lower than those measured at Mauna Loa, while Scandinavian data (obtained aboard an aircraft) and Alaska values are higher, reflecting the greater CO_2 emissions from the more abundant fossil fuel sources in the northern hemisphere.

Major sinks for atmospheric CO_2 are the oceans and the biomass, and since the uptake is rather slow typical residence times are of the order of 2–5 years. Since the beginning of this century about half of the CO_2 produced by burning fossil fuels has been absorbed by the oceans, with a much smaller part going into biomass. The other 50 per cent of the CO_2 released has been accumulating in the atmosphere.

CO_2 reacts with sea water to form a solution of carbonic acid (H_2CO_3) in the form of carbonate (CO_3^{2-}) and bicarbonate ions (HCO_3^-) and hydrogen ions (H^+). The rate of absorption of CO_2 by the oceans depends partly on the water temperature, being greater at lower temperatures. Therefore the main sinks of CO_2 in the ocean are the regions of cold ocean currents. Theoretical considerations by Pytkowicz (1972) suggest that the

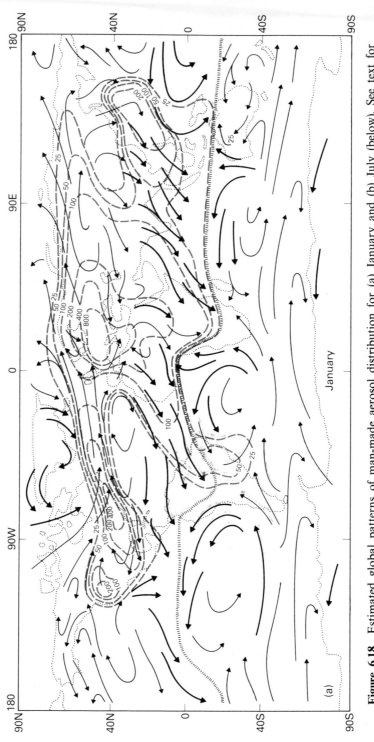

Figure 6.18 Estimated global patterns of man-made aerosol distribution for (a) January and (b) July (below). See text for explanation (*after Kellogg et al., 1975*).

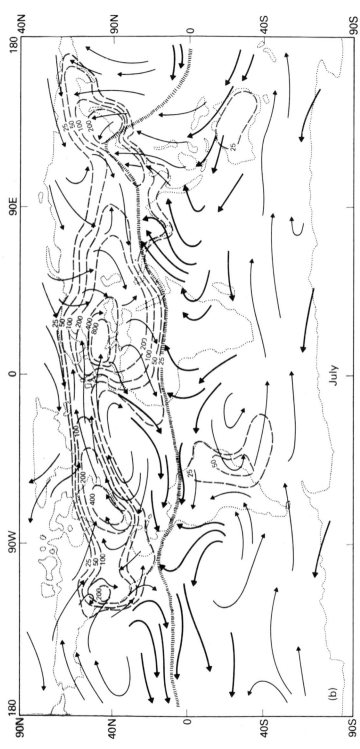

Plate 17   Anthropogenic air pollution from an industrial plant near Johnstown, Pennsylvania. *Reproduced by permission of Grant Heilman, Lititz, PA.*

increasing CO_2 concentration in the atmosphere decreases the pH of the mixed ocean layer above the thermocline, which in turn decreases the capacity of the oceans to absorb the excess CO_2 from fossil fuel combustion. This means that in future a greater proportion of the CO_2 emissions may remain airborne.

Nineteenth century pre-industrial levels of atmospheric CO_2 are estimated by many workers to have been about 290 ppm. Bach (1976) assuming

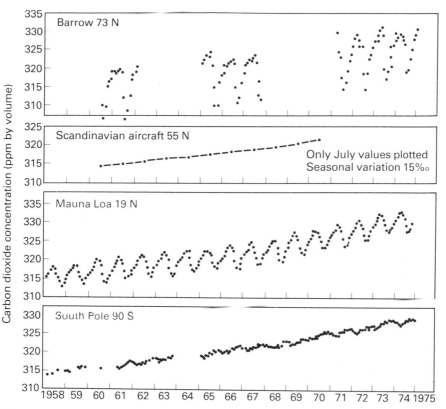

Figure 6.19 Changes in the atmospheric concentration of carbon dioxide (*after Rotty and Weinberg, 1977*).

a fossil fuel consumption growth rate of 4 per cent per year until 1980 and then 3·5 per cent per year thereafter with an average atmospheric retention of 60 per cent of all future CO_2 emissions, has predicted for the year 2000 an atmospheric CO_2 concentration of 385 ppm which is 32 per cent above the base value of 290 ppm. The decline in rate of atmospheric CO_2 increase in the mid-1960's shown in Figure 6.19 occurred during a period of rising industrial CO_2 production, and suggests a temporary increased uptake by either the oceans or land plants. The former possibility was suggested by Bainbridge (1971) as a consequence of the cooling of surface ocean water by about 1 degree C. In 1964 such a cooling occurred in the North Pacific (Namias 1970) and since the cooling trend began almost immediately after the spread of an enormous stratospheric dust cloud from the eruption of the volcano Agung in Bali, a prolonged period of cooler water may account for the declining CO_2 rate.

Bolin (1977) considers that the oceans may have served as a more effective sink for CO_2 than previously considered. He considers that the airborne fraction has been 40 ± 5 per cent, rather than 50 per cent as estimated previously. According to Bolin, the total accumulated input of carbon to the atmosphere since the early nineteenth century has been $210 \pm 30 \times 10^9$ tons. Assuming that the airborne fraction has remained unchanged throughout this period, the amount of carbon in the atmosphere has increased by $90 \pm 15 \times 10^9$ tons; that is to say, the CO_2 concentration has increased by 35 to 50 ppm. Therefore the concentration may have been as low as 275 ppm during the early nineteenth century. Attempts to predict the future increase in the CO_2 content of the atmosphere have been based on the assumption that the airborne fraction of the net output due to human activities has been about 50 per cent during the last 20 years. If, instead, it is 40 ± 5 per cent, the future increase might be slower and possible secondary effects such as climatic changes might be delayed.

3.1.3 *Climatic effects of increasing carbon dioxide*
It was concluded earlier that a continuation of the present fossil fuel consumption growth rate could lead in the year 2000 to atmospheric CO_2 levels about 32 per cent above the pre-industrial base value of 275–290 ppm. Assuming a Gaussian fuel consumption cycle, Hoffert (1974) predicts a doubling of global CO_2 by the year 2025 and a fivefold increase over the pre-industrial value if all fossil fuels are burnt. It is therefore interesting to determine (Table 6.10) what would be the effect of doubling the atmospheric CO_2 content on temperature.

All the authors quoted in Table 6.10 agree that increases of CO_2 result in a warming of the lower atmosphere, but the suggested temperature changes are diverse, ranging from 0·1 to 4 degree C. Among the more interesting results are those of Manabe and Wetherald (1975) obtained by using a three-dimensional general circulation model. Figure 6.20 shows the difference in zonal mean temperature between their $2 \times CO_2$ case and the standard case. Owing to the increase in the greenhouse effect resulting from the increase in the concentration of CO_2, there is a general warming in the model troposphere. In contrast, a large cooling occurs in the model stratosphere. The tropospheric warming is most pronounced in the lower troposphere at high latitudes, and is associated with a decrease in the area of snow and ice. In the model tropics, the warming spreads throughout the entire troposphere due to the intense moist convection, hence its magnitude is relatively small as compared with the warming in the polar region. The greater downward flux of terrestrial radiation resulting from the larger concentration of CO_2 increases the heat energy available for evaporation from the earth's surface, and thus enhances the intensity of the hydrological cycle in the model atmosphere by about 7 per cent. The effect is similar to that discussed for water vapour in the section on the greenhouse effects in Chapter 2.

According to Bryson (1974), another important aspect of the increasing atmospheric CO_2 content is its ability to change the static stability of the atmosphere. While the effect of the changing CO_2 is zero at the tropopause, it is about 0·01 degree C/ppm at the surface. This is enough to change the lapse rate by about 0·001 degree C/km/ppm of CO_2, which, in turn, will modify the latitude of the subtropical anticyclones and the Hadley circulation, since both in part depend on the vertical stability.

Table 6.10 Computed surface temperature changes in degrees Kelvin for given input variations (*after Bach, 1976*).

Model of	Doubling CO_2	Doubling of aerosol optical thickness	Input variations — 1% decrease in solar constant	Input variations — 1% increase in solar constant	Input variations — 1% increase in albedo of earth
Plass (1956)	3·8				
Kaplan (1960)	1·8				
Sawyer (1961, 1966)[a]					
Rakipova (1966)[a]			−0·75 (equator)		
			−0·60 (poles)		
			−0·20 (50°–90° latitude)		
			−0·3 (near equator)		
			−1·4 (near poles)		
Rodgers and Walshaw (1966)	1·95				
Manabe and Wetherald (1967)	2·36[b]	−1·3	−1·3	1·3	
Manabe and Wetherald (1975)[c]	1·33[d]				
	2·93				
Budyko (1969, 1972)	>4		−5		−2
Manabe (1970, 1971)	2·3				
	1·9				
SMIC report (1971)	2·0	−1·8	−1·5	1·5	2·3
Rasool and Schneider (1971)	0·8				
Yamamoto and Tanaka (1972)		−1·3 to −1·8[e]			
		−1·0 to −1·2[f]			
Sellers (1973)	0·1 (global average)				
	0·6 (60°–90° latitude)	<−5	−5·5	0·9	
Paltridge (1974)			0·35		
Weare and Snell (1974)	0·7	−1·1	−0·7		
Smagorinsky (1974)	0·7			0·7	
	1·9[g]			0·6[g]	
	2·9[h]			1·2[h]	
				1·6 (global average)[i]	
				1 to 2 (0°–60° latitude)[i]	
Reck (1975)		0·1 (85° N)		2 to 5 (60°–90° latitude)[i]	
		1·1 (85° S)			

[a] This reference was cited by Robinson (1970).   [b] This value is at fixed relative humidity.
[c] The reader is cautioned not to attach too much significance to the quantitative aspect of the data obtained by Manabe and Wetherald (1975).   [e] Values are for an imaginary refractive index of 0·01.
[d] This value is at fixed absolute humidity.   [f] Values are for an imaginary refractive index of 0·02.   [g] Values are at fixed absolute humidity.
[h] Values are from a one-dimensional model with a fixed relative humidity.
Values are from a three-dimensional model with a fixed absolute humidity.   [i] Values are from a three-dimensional dynamic model.

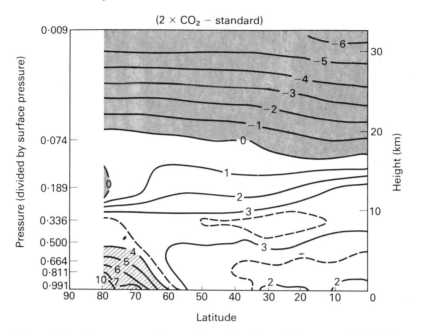

Figure 6.20   Temperature change after a doubling of the atmospheric CO_2 content (*after Manabe and Wetherall, 1975*).

3.2 Anthropogenic heat sources

The magnitudes of anthropogenic heat releases are illustrated in Tables 6.11 and 6.12. Table 6.11 shows the energy consumption density in watts per square metre (obtained by relating the annual energy production to the area over which it is produced) for different locations in comparison with the average net radiation appropriate for the particular location. For entities such as the world or the United States, the heat released is very small and reaches only 0·002–0·5 per cent of the net radiation. In contrast, for urbanized regions having areas of the order of 10^3–10^4 km^2 the artificial heat share reaches 10–30 per cent of the net radiation. Cities at high latitudes (for example, Fairbanks) and those with a compact layout (for example, New York City and Moscow) either already equalize or even exceed considerably the local natural radiation receipt. At the extreme end of the scale, industrial complexes and especially fossil and nuclear power plants are excessive heat islands. Bach (1976) has calculated in Table 6.12 projected energy consumption densities for 1980. It is seen that western Europe and Japan have the highest heat production over their areas and that by 1980 the heat released will increase by about ·55 and 105 per cent, respectively. Recent trends in energy use may change these estimates.

Bach, among several others, has considered to what extent projected global heat releases can conceivably influence climate. The solar input at the top of the atmosphere is about 1360 W m^{-2}. Going from a cross section to a spherical surface reduces this amount by one fourth to 340 W m^{-2}. For

Table 6.11 Comparison of large-scale regional and small-scale densities of energy consumption (*after Bach, 1976*).

Location	Area (km²)	Energy consumption density (W m⁻²)	Average net radiation (W m⁻²)	Per cent energy consumption density of net radiation
World, whole globe	510,000,000	0·016	80–100	0·02–0·016
World, continents only	135,781,867	0·046	80–100	0·05–0·04
United States (without Alaska)	7,827,000	0·26	50–110	0·5–0·2
Central western Europe	1,665,000	0·74	50–60	1·5–1·2
United States (14 eastern states)	932,000	1·11	~80	1·4
Germany (federal republic)	246,000	1·36	~50	2·7
United States (Washington–Boston area)	87,000	4·4	~90	4·9
Northrhein (Westfalia)	34,039	4·2	~50	8·4
Northrhein (industrial area only)	10,296	10·2	~51	20·0
Los Angeles County	10,000	7·5	~108	6·9
Los Angeles (city)	3,500	21	~108	19·5
Cincinnati (standard metropolitan statistical area) ·	1,890	2·8	~99	2·8
Cincinnati (city)	200	26·2	~99	26·5
Moscow	878	127	~42	302
West Berlin	234	21·3	~57	37·4
New York City (Manhattan)	59	630	~93	677
Sheffield	48	19·3	~56	34·5
Fairbanks	37	18·5	~18	100
Refining facility	0·2	440		
Steel mills	0·2	530		
Fossil fuel power plant (1,500 MW)	0·2	10,000		
Nuclear power plant (1,500 MW)	0·2	16,000		

Table 6.12 Energy consumption density by major regions for 1968 and projected to 1980 (*after Bach, 1976*).

	Area (10³ km²)	1968 Energy consumption* (10¹² Btu/yr)	1968 Energy consumption density (W m⁻²)	1980 Energy consumption* (10¹² Btu/yr)	1980 Energy consumption density (W m⁻²)
North America	17,720	68,594	0·13	106,124	0·20
United States (without Alaska	7,827	62,432	0·27	95,145	0·41
Canada	9,960	6,162	0·02	10,979	0·04
Western Europe	1,665	41,584	0·03	64,354	1·29
USSR and eastern Europe	23,644	39,843	0·06	80,073	0·11
USSR	22,403	28,628	0·04	60,611	0·09
Eastern Europe	1,241	11,215	0·30	19,462	0·52
Non-communist Asia	17,028	16,757	0·03	38,666	0·08
Japan	369	8,691	0·79	17,715	1·61
Other Asian countries	16,659	8,066	0·02	20,951	0·04
Communist Asia	10,021	9,342	0·03	25,690	0·09
Africa	30,284	3,343	0·004	7,236	0·008
Latin America	20,096	8,034	0·01	18,771	0·03
Australia and Oceania	8,557	2,240	0.009	3,743	0·02
World (continents)	135,781	189,737	0·047	344,657	0·085

* Energy consumption values are from Darmstadter (1973).

a planetary albedo of 0·3, about 240 W m⁻² are retained, of which about 65 per cent is absorbed by the surface of the earth, resulting in a value of 150 W m⁻² for the whole globe or 100 W m⁻² if related only to continents. According to Weinberg and Hammond (1970) the average *per capita* energy consumption in the United States was about 10^4 W in 1970. Bach assumes an upper limit to total world population of 20×10^9 (in 2000 AD it is forecast to be $6·5 \times 10^9$), with a *per capita* income four times the present US income, and by scaling the present *per capita* power consumption by the same factor, obtains 800 TW (10^{12} W or 1·57 W m⁻²) as an upper limit to global energy production. This is an increase by a factor of 100 over the 1970 global anthropogenic heat emission of 8 TW.

   Let *H* denote the total global heat input, then a fractional increase $\Delta H/H$ can be related to a mean temperature change ΔT. Climatic models such as those already described in Chapter 3 by Budyko (1969) and Sellers (1969, 1973) predict a temperature change of 2–3 degree C for a 1 per cent change in heat input. The model results show also that the changes are twice as large near the poles. Relating now the anthropogenic heat contribution of 800 TW (1·57 W m⁻²) to the global solar heat absorption for the whole globe (150 W m⁻²) a fractional increase $\Delta H/H$ of 1 per cent is obtained. Thus the total additional heat from man's activities could lead to an average temperature increase of 2–3 degree C for the globe and perhaps 10 degree C for the polar regions. Possible effects of such a temperature increase would include the melting of polar glaciers and snow fields.

Model estimates of this type are useful in that they illustrate the magnitude of the problem. The actual estimates must be greatly in error, but suggest that anthropogenic heat is unlikely to influence global climate in the immediate future, though it may be a very-long-term problem.

While anthropogenic heat sources do not at present influence global climates, they can cause observable effects on local climates. The heat generated in a large city creates a so-called 'urban heat island', which is a small (a few degrees) thermal 'bump' on the lowest few hundred metres of the planetary boundary layer of the atmosphere. The influence of urbanization on local climates is discussed in a later section.

Hanna and Gifford (1975) have listed the energy characteristics of natural and man-made sources in Table 6.13. The heat generated in a large city is

Table 6.13   Energy characteristics of natural and man-made sources (*after Hanna and Gifford, 1975*).

Source	Area	Time duration	Total power	Observations
Surtsey volcano	1 km²	Several months	100,000 MW	Continuous cloud water spouts
Australian bush fire	50 km²	Several hours	100,000 MW	Cumulus cloud convergence
Booster rocket test	300 m²	150 seconds	148,000 MW	Cumulus cloud
Oil burners	3·2 km²	Several hours	700 MW	Cumulus cloud dust devils
Oil fires	—	Day	10,000 MW	Large plume
Large city	10³ km²	Continuous	100,000 MW	Effects on climate
Thunderstorm	10 km²	Hour	50,000 MW	2 cm/hr rain
Power park	5 to 100 km²	Continuous	100,000 MW	—

spread over a wide area, but some of the other anthropogenic heat sources are local and intense, and generate heat plumes which penetrate the boundary layer and rise to great elevations in the troposphere. One of the most intense anthropogenic heat sources is the electric-power generating station. The maximum amount of electric power currently generated at a single thermal power station site is about 3,000 MW. Hanna and Gifford conclude that the atmospheric effects of heat dissipation from current power stations are not serious, provided that efforts are made to design the facility such that downwash is eliminated, drift is minimized, and plume rise is maximized. Clouds are sometimes observed to form due to heat and moisture from cooling towers, but no significant changes in rainfall in the areas of study have been detected.

However, Hanna and Gifford report that several sites are being studied by American power utility companies and US government agencies as potential 'power parks' or energy centres. Such fossil or nuclear-powered plants could generate 10,000 to 50,000 MW of usable power on a land area of the order of 5 to 100 km². Assuming 33 per cent efficiency, there might be as much as 100,000 MW of waste energy dissipated to the atmosphere, probably through some type of cooling towers. This dissipation rate represents an increase of between one and two orders of magnitude of the energy released over that from the largest currently-operating power plants, and the meteorological effects of such a power input to the atmosphere are uncertain. The only comparable stationary heat sources of this magnitude are large fires and certain geophysical phenomena. Bourne (1964) and Thorarinsson and Vonnegut (1964) described the meteorological phenomena that accompanied the Surtsey volcano, which released an estimated 100,000 MW of heat continuously to the atmosphere from an area less than 1 km². This energy was released in the form of sensible (convective) heat and was accompanied by the release of many small ash particles. A permanent cloud extending to heights of 5 to 9 km, visible 115 km away, formed over the volcano. Waterspouts formed below the bent-over plume from the volcano, indicating that the indraft at the cloud base acted to concentrate local atmospheric vorticity. Observations of volcanos and brush fires suggests that the proposed power parks may cause the development of large clouds, and may sometimes trigger thunderstorm and whirlwind activity in the area.

3.3 Land-surface alterations

The most widespread modification of the climate by man in the past has been achieved inadvertently by the conversion of the natural vegetation into arable land and pastures (SMIC report 1971). According to Flohn (1973), during the past 8,000 years, about 11 per cent of the land area has been converted to arable land, and 31 per cent of the forest land is no longer in its natural state. The modification of the natural vegetation affects several significant climatic parameters such as, the surface roughness, the surface albedo, and the Bowen ratio.

It is well established that the principal effects of forest on the hydrological cycle are in the reception and disposal of precipitation (Pereira 1972). Infiltration of rainwater or of snowmelt in the ground under forest is usually greater than under any alternative form of land use. Forests also provide the greatest surface area for the interception and re-evaporation of water and are effective traps for the absorption of solar radiation. Tree root depths are characteristically large, so trees are frequently not under moisture stress at times when grasslands have dried up.

In the high rainfall middle latitudes, grasslands occupy much of the area whose ecological climax vegetation is forest. Temperate grass has become a crop of major importance and is managed for high productivity. Hydrologically this intensively-managed grassland gives, according to Pereira, efficient infiltration of rainfall, excellent soil protection and very low run-off, while water use is heavy. The grass thus exhibits, hydrologically, much of the regulating and protective character of the forest which it has replaced. It is, however, characteristically shallow rooted in comparison with forest and therefore its water use is less.

The steppes, savannas and prairies of the great continents lie in ecological conditions which do not sustain forests. They have long dry seasons and over vast areas of the tropics and subtropics they are semi-arid. The effects of the deterioration of the vegetation cover on the climate of semi-arid regions were discussed in Chapter 3.

The growing of crops for food, fibre or vegetable oils involves the complete destruction of the natural vegetation and exposure of the mineral soil. The effect of clearing the land of vegetation is to reduce the total water use and to leave more water available, either to percolate deeply to groundwater or else to run-off over the surface.

In Table 6.14, the SMIC report (1971) estimates the changes that take place after the conversion of a temperate-zone forest to agricultural fields. An effective incoming short-wave radiation of 125 W m^{-2} and an outgoing radiation of 50 W m^{-2} over the fields, are assumed in the report. Changing from forest to dry arable land reduces the evaporation and increases the sensible heat flux. In contrast, if the arable land is irrigated the evaporation will increase.

Table 6.14 Changes of heat budget after conversion from forest to agricultural use (*data from SMIC Report, 1971*).

	Albedo	Net radiation (W m^{-2})	Sensible heat-flux (W m^{-2})	Latent heat-flux (W m^{-2})	Bowen ratio	Evapo-trans-piration, mm per month
Coniferous forest	0·12	60	20	40	0·50	41
Deciduous forest	0·18	53	13	39	0·33	40
Arable land, wet	0·20	50	8	42	0·19	43
Arable land, dry	0·20	50	15	35	0·41	36
Grassland	0·20	50	20	30	0·67	31

Irrigation has been among man's major interventions in the hydrological cycle. Estimates of global water consumption are contained in Table 6.15. The 1,700 km^3 per year of non-returned irrigation water represents a change of about 5 per cent in the run-off of the land areas. The 1,700 km^3 per year additional evaporation is approximately 2 per cent of the total annual evaporation of the land areas. The SMIC report (1971) compared the energy required to evaporate this irrigation water with the energy introduced into his environment by man's use of power. To evaporate 1,700 km^3 of water requires over 100 TW, while man's power production now amounts to about 8 TW. The direct influence of the energy used for the evaporation of irrigation water is to speed the hydrological cycle, since all the heat used in evaporation is returned to the atmosphere where condensation occurs. In contrast, man's energy production enters directly into the overall heat budget of the earth.

Irrigation can have both local and global effects. Flohn (1972) has described the water balance of old Saharan oases in southern Tunisia, and his

Table 6.15 Estimated global water needs and water resources (km³ per year) (*after Lvovich, 1969*).

	1965		2000	
	Withdrawal	Return	Withdrawal	Return
Needs:				
Municipal water supply	98	56	950	760
Irrigation	2300	600	4250	400
Industry	200	160	3000	2400
Energy	250	235	4500	4230
	2848	1051	12,700	7790
Resources:				
Precipitation on continents		108,000		
Run-off		37,000		
Evapotranspiration		71,000		
Evapotranspiration from agricultural areas		(3560)		

results are shown in Table 6.16. The evaporation from the humid oases is high and utilizes energy from the air giving rise to a negative Bowen ratio. The Bowen ratio of the semi-desert is large and positive, suggesting limited evaporation and a high sensible heat flux. The total agricultural production of the oases has increased since 1881 by at least 50 per cent, and Flohn assumes a similar rise in the water consumption. Apart from irregular fluctuations and local deviations, the area-averaged annual rainfall is constant, and amounts to 148 mm. Assuming that precipitation and evaporation were equal at the beginning of the modern era (1881), the recent additional water consumption (or evaporation) of the oases amounts to about 2·5 mm per year for the total area (35,000 km²) or nearly 2 per cent. This additional water consumption is supplied from artesian water, which consists partly of fossil water and this will not be replenished under the present climatic conditions.

Table 6.16 suggests that locally the air over oases is being cooled, the energy loss being used to evaporate irrigation water. Now an irrigated area

Table 6.16 Water balance of Saharan oases in southern Tunisia (*after Flohn, 1972*).

Area (km²)	Oases average 150	Douz (humid) 0·63	El Hammed Gabes (dry) 11·6	Semi-desert 35,000
Albedo (assumed)	0·15	0·15	0·15	0·25
Net radiation (Ly/day)	200	212	192	158
Net radiation equivalent ET (cm/yr)	124	131	119	98
Water consumption (cm/yr)	153	346	78	—
Precipitation P (cm/yr)	15	8	16	15
Observed ET (cm/yr)	168	354	94	15
Observed ET (Ly/day)	272	572	152	24
Sensible heat flux (Ly/day)	−72	−360	−40	134
Bowen ratio	−0·26	−0·63	0·26	5·6

grows vegetation that has an albedo significantly lower than the semi-arid ground cover that has been replaced. The interesting consequence of this is that while the local temperature is lowered by irrigation because of the increased evaporation, global temperature is raised because of the decreased reflection of incoming solar radiation. Budyko's (1974) calculation that present-day irrigation leads to an increase of the earth's mean surface temperature by about 0·07 degree C is derived from a consideration of this effect.

3.3.1 *Urbanization*

There is no doubt that whenever man changes the landscape he modifies the microclimate (SMIC report 1971). One of the greatest modifications is the building of large towns and cities. When groups of buildings merge with towns and cities, they generate a 'climatological dome' with a set of well documented meteorological anomalies. Frequently the top of this dome is well defined and can be identified quite readily from an aircraft.

According to the SMIC report, cities modify the local climate in two particular ways. Firstly, an urban thermal and pollution plume exists whenever a wind is blowing, transporting heat and matter downstream out of the city and modifying, for example, the rural radiation balance for at least a few kilometres. Secondly, a feature of some interest occurs whenever the regional flow is weak. Thermal gradients generated within the city set the air in motion and urban wind cells develop as illustrated schematically in Figure 6.21. The thermal gradients are largely a result of heat released into the atmosphere by industrial, transportation and domestic sources within the densely populated urban areas.

The main effect of this heat output on the local scale is the creation of a stationary three-dimensional heat island, as described, for example, by Chandler (1965) for London. The intensity of this heat island reaches a maximum during the night, when the surrounding rural areas are cooling under the effect of outgoing thermal radiation. The minimum intensity occurs

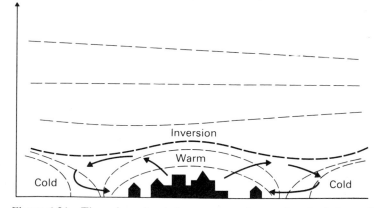

Figure 6.21 The urban circulation pattern when regional winds are light.

shortly after noon, when the sensible heat flux of the surrounding areas more or less equals the input of artificial heat. Under stable conditions with light winds, the effect is restricted to a shallow atmospheric layer a few hundred metres thick, the surface warming being about 2 to 6 degree C.

According to the SMIC report, additional urban climatic changes are caused by the increased surface roughness, the changed albedo, the accelerated run-off, and the changed heat-storage capacity resulting from the replacement of forests and fields by concrete and buildings. The SMIC report does not consider that the influence of these latter processes will extend significantly beyond the built-up areas.

Urban heat sources can cause local de-stabilization of the surface layer and lead to an increase of frequency of showers and thunderstorms. Claasen (1970) has observed a local increase in shower frequency at the southern downwind border of the industrial region of western Ruhrgebiet, West Germany.

3.4 Trends in global water balance

Estimated global water needs in 1965 and 2000 AD, and global water resources are shown in Table 6.15. The increased water used by man for domestic and industrial purposes and even larger use for irrigation leads to an acceleration of the hydrological cycle over land. According to the SMIC report (1971), the intensity of this process depends on two quantities: the additional non-returnable water consumption (decrease in run-off) per head V_a, and the population density P. If V_a is given in litres per day per person and P in persons/km^2, then the additional evaporation E_a is given by

$$E_a = V_a P(10^{-7} \text{ cm/day})$$

A good example is provided by Keller's (1972) study of the Federal Republic of Germany. He compared the period 1931 to 1960 with the period 1891 to 1930, and obtained an increase of precipitation of 3 per cent, a decrease of run-off of 12 per cent, and, as a residual, an increase of evapotranspiration of about 20 per cent, or 80 mm per year. These changes can be attributed partly to a rise in agricultural production and industrial and domestic water demand.

The SMIC report takes Lvovich's (1969) value for non-returnable water consumption of 1,800 km^3 per year for 1965, and obtains a value for V_a of about 1,500 litres per day as a global average, of which more than 90 per cent is used for irrigation. Using the preceding formula for western and central Europe ($P = 169$/km^2), a value of 9 cm for E_a is obtained, co-inciding well with Keller's result.

Estimates of global water uses are summarized in Table 6.15 by Lvovich. As the differences between the withdrawals and returns indicate, irrigation is by far the most significant consumption, causing about a 5 per cent decrease in the natural run-off. By the year 2000, Lvovich estimates that irrigation will cause a decrease in global run-off of about 10 per cent. The total decrease in global run-off in the year 2000 could reach 13 per cent.

Changes are also taking place in the groundwater reservoirs of the earth. These were filled up gradually during the geological history, and there is evidence that this process is now being reversed by extraction rates of water surpassing the rates of natural re-charge. Simultaneous increases in the amounts of water in the world ocean and the polar ice masses observed during the last 80 years may be partially attributed to groundwater extractions.

References

BACH, W. 1976: Global air pollution and climatic change. *Reviews Geophysics Space Physics* **14**, 429–74.

BACH, W. and DANIELS, A. 1975: *Handbook of air quality in the United States.* Honolulu: Oriental Publishing Company.

BAINBRIDGE, A. E. 1971: Atmospheric carbon dioxide variations. *Transactions American Geophysical Union* **52**, 222.

BOLIN, B. 1977: Changes of land biota and their importance for the carbon cycle. *Science* **196**, 613–15.

BOURNE, A. G. 1964: Birth of an island. *Discovery* **25**, 16–19.

BROWN, L. R. 1975: *The global politics of food: role and responsibility of North America.* Toronto: University of Toronto, Masfleet-Falconer Lecture.

BRYSON, R. A. 1974: A perspective on climatic change. *Science* **184**, 753–60.

BUDYKO, M. I. 1969: The effect of solar radiation variations on the climate of the earth. *Tellus* **21**, 611–19.

1972: The future climate. *EOS Transactions, AGU* **53**, 868–74.

1974: *Climate and life.* New York: Academic Press.

CHANDLER, T. J. 1965: *The climate of London.* London: Hutchinson.

CHARNEY, J. G., STONE, P. H. and QUIRK, W. J. 1976: Drought in the Sahara: insufficient biogeophysical feedback? *Science* **191**, 100–2.

CLAASON, C. 1970: *Untersuchungen über die Haufigkeitsverstarkung von Niederschlagseclos im 100 km—Umkreis um Bonn mittels 3 cm Radar.* Bonn: Diplomarbeit, University of Bonn.

COBB, W. E. 1973: Oceanic aerosol levels deduced from measurements of the electrical conductivity of the atmosphere. *Journal Atmospheric Science* **30**, 100–6.

COBB, W. E. and WELLS, H. I. 1970: The electrical conductivity of oceanic air and its correlation to global atmospheric pollution. *Journal Atmospheric Sciences* **27**, 814–19.

COLLIS, R. T. H. 1975; Weather and world food. *Bulletin American Meteorological Society* **56**, 1078–83.

CRONIN, J. F. 1971: Recent volcanism and the stratosphere. *Science* **172**, 847–9.

DARMSTADTER J. 1973: Energy consumption: trends and patterns. In SCHURR, S. H., editor, *Energy, economic growth and environment*, 155–223. Baltimore: Johns Hopkins University Press.

ELLIS, H. T. and PUESCHEL, R. F. 1971: Absence of air pollution trends at Mauna Loa. *Science* **172**, 845–6.

FAIRBRIDGE, R. W. 1976: Effects of holocene climatic change on some tropical geomorphic processes. *Quaternary Research* **6**, 529–56.

FEDOSEEV, A. P. 1958: Agrometeorological conditions for the development of pasture yields and the principles of their estimation and prognostication. In DAVITAYA, F. F. and KULIK, M. S., editors, *Compendium of abridged reports to the second session of CAgM (WMO)*, 54–61. Moscow: Hydrometeorological Publishing House.

FISCHER, W. H. 1967: Some atmospheric turbidity measurements in Antarctica. *Journal Applied Meteorology* **6**, 958–9.

FLOHN, H. 1972: The influence of man on the hydrological cycle: discussion. In IASH, *World water balance* **3**, 518. Gentbrugge.

— 1973: Globale energiebilanz und klimaschwankungen. In *Bonner Meteorologische Abhandlungen*, 75–117. Westdeutscher Verlag.

GOH, K. C. 1975: *The influence of topographic and synoptic factors on rainfall distributions in the central Pennines*. Leeds: unpublished PhD thesis, University of Leeds.

GOH, K. C. and LOCKWOOD, J. G. 1974: An assessment of topographical controls on the distribution of rainfall in the central Pennines. *Meteorological Magazine* **103**, 275–87.

GRINDLEY, J. 1972: Estimation and mapping of evaporation. In IASH, *World water balance* **1**, 200–213. Gentbrugge.

HANNA, S. R. and GIFFORD, F. A. 1975: Meteorological effects of energy dissipation at large power parks. *Bulletin American Meteorological Society* **56**, 1069–76.

HIDY, G. M. and BROCK, J. R. 1971: An assessment of the global sources of tropospheric aerosols. In *Proceedings of the 2nd International Clean Air Congress*, 1088–97. New York: Academic Press.

HOFFERT, M. I. 1974: Global distributions of atmospheric CO_2 in the fossil-fuel era: a projection. *Atmospheric Environment* **8**, 1225–49.

HOUNAM, C. E., BURGOS, J. J., KALIK, M. S., PALMER, W. C. and RODDA, J. 1975: *Drought and agriculture*. WMO technical note, **38**. Geneva.

KAPLAN, L. D. 1960: The influence of carbon dioxide variations on the atmospheric heat balance. *Tellus* **12**, 204–8.

KELLER, R. 1972: Water-balance in the Federal-Republic of Germany. In IASH, *World Water Balance* **2**, 300–14. Gentbrugge.

KELLOGG, W. W., COAKLEY, J. A. JR. and GRAMS, G. W. 1975: Effect of anthropogenic aerosols on the global climate. In *Proceedings of the WMO/IAMAP symposium on long-term climatic fluctuations*, 323–30. Geneva: WMO.

KNOX, J. B. and MACCRACKEN, M. C. 1976: Concerning possible effects of air pollution on climate. *Bulletin American Meteorological Society* **57**, 988–91.

KUTZBACH, J. 1974: Fluctuations of climate—monitoring and modelling. *WMO Bulletin* **23**, 155–63.

LANDSBERG, H. E. 1976: Whence global climate: hot or cold? an essay review. *Bulletin American Meteorological Society* **57**, 441–3.

LETTAU, H. 1969: Evapotranspiration climatonomy. 1. A new approach to numerical prediction of monthly evapotranspiration, runoff, and soil moisture storage. *Monthly Weather Review* **97**, 691–9.

LETTAU, H. and LETTAU, K. 1969: Shortwave radiation climatonomy. *Tellus* **21**, 208–22.

LOCKWOOD, J. G. and VENKATASAWMY, K. 1975: Evapotranspiration and soil moisture in upland grass catchments in the eastern Pennines. *Journal Hydrology* **26**, 79–94.

LUDWIG, J. H. MORGAN, G. B. and MCMULLEN, T. B. 1970: Trends in urban air quality. *Eos Transactions AGU* **51**, 468–75.

LUDWIG, J. H., MORGAN, G. B. and MCMULLEN, T. B. 1971: Trends in urban air quality. In MATTHEWS, W. H., KELLOGG, W. W. and ROBINSON, G. D., editors, *Man's impact on the climate*, 321–38. Cambridge, Mass.: MIT Press.

LVOVICH, M. I. 1969: *Vednie Resourse Budushevo (Water resources of the future)*. Moscow: Prosveschenie.

MANABE, S. 1970: The dependence of atmospheric temperature on the concentration of carbon dioxide. In SINGER, S. F., editor, *Global effects of environmental pollution*, 25–9. New York: Springer.

1971: Estimates of future change of climate due to the increase of carbon dioxide concentration in the air. In MATTHEWS, W. H., KELLOGG, W. W. and ROBINSON, G. D., editors, *Man's impact on climate*, 249–64. Cambridge, Mass.: MIT Press.

MANABE, S. and WETHERALD, R. T. 1967: Thermal equilibrium of the atmosphere with a given distribution of relative humidity. *Journal Atmospheric Sciences* **24**, 241–59.

1975: The effects of doubling the CO_2 concentration on the climate of a general circulation model. *Journal Atmospheric Sciences* **32**, 3–15.

MASON, B. J. 1976: Towards the understanding and prediction of climatic variations. *Quarterly Journal Royal Meteorological Society* **102**, 473–98.

MCQUIGG, J. D. 1974: *Weather and climate interactions with grain yields, 1975 agricultural outlook*. Washington: US Government Printing Office, Ninety-third Congress, 140–4.

MURRAY, R. 1977: The 1975/76 drought over the United Kingdom—hydrometeorological aspects. *Meteorological Magazine* **106**, 129–45.

NAMIAS, J. 1970: Macroscale variations in sea-surface temperatures in the North Pacific. *Journal Geophysical Research* **75**, 565–82.

NATIONAL ACADEMY OF SCIENCES 1976: *Climate and food: climatic fluctuation and US agricultural production*. Washington.

O'KEEFE, P. and WISNER, B. 1975: African drought—the state of the game. In RICHARDS, P., editor, *African environment: problems and perspectives*, 31–9. London: International African Institute.

PALTRIDGE, E. W. 1974: Global cloud cover and earth surface temperature. *Journal Atmospheric Science* **31**, 1571–6.

PENMAN, H. L. 1948: Natural evaporation from open water, bare soil and grass. *Proceedings Royal Society* Series A, **193**, 120–45.

1963: *Vegetation and hydrology*. Commonwealth Bureau of Soils technical communication, **53**. Harpenden.

PEREIRA, H. C. 1972: The influence of man on the hydrological cycle. In IASH, *World water balance* **3**, 553–69. Gentbrugge.

PETERSON, J. T. and JUNGE, C. E. 1971: Sources of particulate matter in the atmosphere. In MATTHEWS, W. H., KELLOGG, W. W. and ROBINSON, G. D. editors, *Man's impact on the climate*, 310–20. Cambridge, Mass.: MIT Press.

PLASS, G. N. 1956: The influence of the 15 carbon dioxide band on the atmospheric infrared cooling rate. *Quarterly Journal Royal Meteorological Society* **82**, 310–24.

PRIESTLEY, C. H. B. and TAYLOR, R. J. 1972: On the assessment of surface heat flux and evaporation using large-scale parameters. *Monthly Weather Review* **100**, 81–92.

PYTKOWICZ, R. M. 1972: Fossil fuel burning and carbon dioxide—a pessimistic view. *Comments Earth Science Geophysics* **3**, 15–22.

RAPP, A. 1974: *A review of desertization in Africa: water, vegetation and man.* Stockholm: Secretariat for International Ecology, Sweden (SIES), report **1**.

RASOOL, S. I. and SCHNEIDER, S. H. 1971: Atmospheric carbon dioxide and aerosols: effects of large increases on global climate. *Science* **173**, 138–41.

RECK, R. A. 1975: Aerosols and polar temperature changes. *Science* **188**, 728–30.

ROBINSON, G. D. 1970: *Long-term effects of air pollution—a survey*, 402–40. Hartford, Conn.: Center for the Environment and Man, Inc.

RODGERS, C. D. and WALSHAW, C. D. 1966: The computation of infrared cooling rate in planetary atmospheres. *Quarterly Journal Royal Meteorological Society* **92**, 67–92.

ROTTY, R. M. and WEINBERG, A. M. 1977: How long is coal's future? *Climatic Change* **1**, 45–57.

SCEP 1970: Report on the study of critical environment problems. In *Man's impact on the global environment.* Cambridge, Mass.: MIT Press.

SCHNEIDER S. H. and MESIROW, L. E. 1976: *The genesis strategy: climate and global survival.* New York: Plenum Press.

SELLERS, W. D. 1969: A global climate model based on the energy balance of the earth–atmosphere system. *Journal Applied Meteorology* **8**, 392–400.
     1973: A new global climatic model *Journal Applied Meteorology* **12**, 241–54.

SHEETS, H. and MORRIS, R. 1974: *Disaster in the desert.* Washington: Carnegie Endowment for International Peace.

SMAGORINSKY, J. 1974: Global atmospheric modelling and the numerical simulation of climate. In HESS, W. N., editor, *Weather and climate modification*, 633–86. New York: John Wiley.

SMIC REPORT 1971: *Inadvertent climate modification. Report of the study of man's impact on climate.* Cambridge, Mass.: MIT Press.

SPREEN, W. C. 1947: A determination of the effect of topography upon precipitation. *Transactions American Geophysical Union* **28**, 285–90.

SURAQUI, S., TABOR, H., KLEIN, W. H. and GOLDBERG, B. 1974: Solar radiation changes at Mt St Katherine after forty years. *Solar Energy* **16**, 155–8.

TABONY, R. C. 1977: Drought classifications and a study of droughts at Kew. *Meteorological Magazine* **106**, 1–10.

THORARINSSON S. and VONNEGUT, B. 1964: Whirlwinds produced by the eruption of Surtsey Volcano. *Bulletin American Meteorological Society* **45**, 440–4.

THORNTHWAITE, C. W. and MATHER, J. R. 1955: The water budget and its use in irrigation. In *Water—the yearbook of agriculture 1955*, 346–58. Washington: US Department of Agriculture.

UN PROTEIN ADVISORY GROUP 1974: *Recommendations. August 1974.* Geneva: United Nations.

WADE, N. 1974: Sahelian drought: no victory for western aid. *Science* **185**, 234–7.

WALES-SMITH, B. G. 1971: Monthly and annual totals of rainfall representative of Kew, Surrey, for 1697–1970. *Meteorological Magazine* **100**, 345–62.
     1973a: An analysis of monthly rainfall totals representative of Kew, Surrey, from 1697–1970. *Meteorological Magazine* **102**, 157–71.

1973b: Evaporation in the London area from 1698 to 1970. *Meteorological Magazine* **102**, 281–91.

WEARNE, B. C. and SNELL, F. M. 1974: A diffuse thin cloud atmospheric structure as a feedback mechanism in global climatic modelling. *Journal Atmospheric Science* **31**, 1725–34.

WEINBERG, A. M. and HAMMOND, R. P. 1970: Limits to the use of energy. *American Scientist* **58**, 412–18.

WIGLEY, T. M. L. and ATKINSON, T. C. 1977: Dry years in south-east England since 1698. *Nature* **265**, 431–4.

WINSTANLEY, D. 1975: The impact of regional climatic fluctuations on man: some global implications. In *Proceedings of the WMO/IAMAP symposium on long term climatic fluctuations*, 479–91. Geneva: WMO.

YAMAMOTO, G. and TANAKA, M. 1972: Increase of global albedo due to air pollution. *Journal Applied Meteorology* **29**, 1405–12.

Conclusion: Man and Climate

The problem of predicting the future development of global climate has been in the limelight for several years. Most meteorologists have approached the problem very cautiously, since they are well aware of the complexity of the atmospheric system. Many of the problems have been discussed in earlier chapters of this book. Flohn (1977) has suggested that essentially three alternatives seem to exist for the probable climatic development during the next century:
  (1) preservation of the present state with variations similar to those observed during the last 200 years;
  (2) transition to another balanced state: that of an ice age;
  (3) overwhelming growth of man-made effects: rapid transition to a warm age.

According to Mitchell (1970) there has been a systematic fluctuation in global climate during the last century characterized by a net world-wide warming of about 0·6 degree C between the 1880's and the early 1940's followed by a net cooling of 0·2 to 0·3 degree C by 1970. These changes are shown for the high-latitude northern hemisphere in Figure 5.8 by a spectral analysis of oxygen-18 values from the Camp Century ice core. The smoothed curve repeats itself at regular intervals and the period 1350–1500 is a good analogue for the present century. So if past trends are followed in the near future, the last part of the present century should be colder than normal in northern latitudes with a slight warming around the year 2000. Damon and Kunen (1976) have concluded that cooling since the early 1940's is not necessarily global in extent; rather, it seems to be largely limited to part of the northern hemisphere and some southern hemisphere sectors such as subequatorial Africa. Cooling is most evident at high northern latitudes, since data for 60° to 80° N account for about one-half of the net secular trend of the hemisphere, despite the small geographical area represented. Sectors of the southern hemisphere south of 45° S show significant temperature trends, which are balanced by opposite changes in other sectors. For example, a cooling trend in subequatorial Africa is offset by a warming trend in Australia and New Zealand. According to Damon and Kunen, fourteen stations between 45° and 90° S for which surface air

Plate 18 Clearing snow in Buffalo, USA, during February 1977, after the worst snow storm for many years. *Reproduced by permission of Keystone Press Agency Ltd.*

temperature data for the four pentads from 1955 to 1974 are available, show a mean decrease of 0·12 degree C from the 1955–9 pentad, followed by an increase of 0·37 degree C to 1970–4 pentad. This apparent warming trend appears to be amplified at higher latitudes.

Global surface temperature depends largely on the solar constant, the

Plate 19   Cars buried in Buffalo, USA, during February 1977, after the worst snow storm for many years. *Reproduced by permission of Keystone Press Agency Ltd.*

amount of volcanic dust in the atmosphere and the atmospheric CO_2 content. Schneider and Mass (1975a, 1975b) have combined the solar energy effect with the effect of volcanic dust in a climate model. They related the solar constant to the number of sunspots using the relationship of Kondratyev and Nikolsky (1970) described in Chapter 2. Volcanic dust veils have been proposed as an external cause of climatic change primarily because they can scatter and absorb several per cent of the direct solar beam, thereby preventing some of the solar energy from reaching the lower atmosphere. This is discussed in detail in Chapters 2 and 6. The 'dust veil index' compiled and tabulated by Lamb (1970) and described in Chapter 2 was used by Schneider and Mass to simulate volcanic eruptions. The resulting climatic model curve is shown in Figure 7.1. The calculated curve is similar in some general features to the historical records discussed in Chapter 5. Particularly striking are the 'Little Ice Age' temperature minimum between 1650 and 1700, the subsequent rise in temperature until about 1800 and the fall and rise to 1880. There is a rise in temperature after 1890 corresponding to a period of negligible volcanic activity between 1915 and 1963 (when Mount Agung erupted). Another interesting feature shown in Figure 7.1 is the rise in computed global temperature beginning about 1968, since there is widespread belief that the earth's surface temperature has been cooling since the 1950's. Thus either: (1) the sunspot–volcanic dust mechanism used by Schneider and Mass is incomplete or wrong; or (ii) the 'cooling trend' is

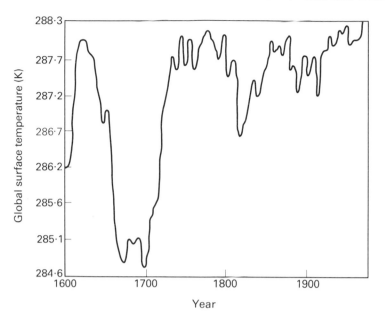

Figure 7.1 Global average surface temperature computed by Schneider and Mass (*1975a, 1975b*) using as input forcing the sunspot data in Figure 2.3 combined with the Kondratyev and Nikolsky solar irradiance–sunspot formula and an equivalent decrease in solar input to the lower atmosphere from the effect of volcanic dust veils as given by Lamb.

not global; or (iii) some other mechanism has been operating in addition to those modelled. It was suggested earlier in this chapter that the evidence for global cooling has been based, in large part, on a severe cooling trend in high northern latitudes.

Carbon dioxide and anthropogenic particulate matter are the two major pollutants not discussed in the Schneider and Mass climatic model that may be affecting current climatic trends. The major source of both of these pollutants is the highly populated and industrialized northern hemisphere. Because of the relatively fast latitudinal mixing rates in the atmosphere, differences in CO_2 concentrations between the northern and southern hemispheres are relatively small. Figure 6.18 shows that anthropogenic particulate pollutants are mostly restricted to the northern hemisphere. Consequently, the climatic effect of CO_2 will manifest itself almost equally in both hemispheres, whereas the effect of anthropogenic particulate matter pollution will be most intense in the northern hemisphere. There is general agreement that the effect of the increasing burden of atmospheric CO_2 will be a global warming, although the exact magnitude of the effect is still debated. However, the direct effect of anthropogenic particulate matter pollution can be either a cooling or warming, depending on the size and distribution of the particles.

Schneider and Mass (1975a, 1975b) have extrapolated sunspot activity to the year 1989 so as to forecast a temperature pattern for a volcano-free period. It is well known that the human input of carbon dioxide into the atmosphere is increasing and this was included in the calculations. Schneider and Mass found (Figure 7.2) that the CO_2 warming dominates the surface temperature pattern soon after 1980, but that superimposed on this warming is a very strong temperature minimum between 1976 and 1981. This is at least similar to the projections from the Camp Century ice core which suggest a temperature minimum towards the end of the present century. Damon and Kunen (1976) comment that CO_2 greenhouse warming trend should first become evident in the southern hemisphere because of the heavy particulate pollution in the northern.

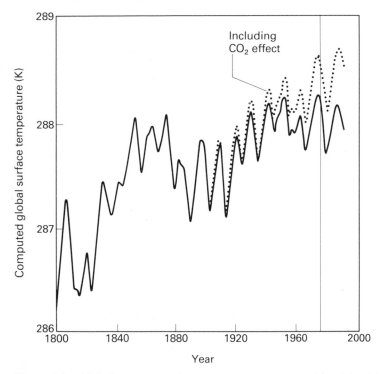

Figure 7.2 Global average surface temperature computed by Schneider and Mass (*1975a, 1975b*). The solid line is for the nominal case given by Figure 7.1, and the dotted curve includes the added effect of increasing atmospheric CO_2. The calculations for 1975 to 1989 are not based on actual observations, but are extrapolations.

CO_2 in the atmosphere affects the thermal radiative balance of the planet and therefore the global climate. Temperature changes due to doubling the CO_2 concentration in the air are shown in Figure 6.20. The cooling of the stratosphere and especially the pronounced warming near the surface in higher latitudes are highly significant. The effects of warming the earth's surface north of 70° N by more than 7 degree C would be extreme. It would lead to a climatic situation which has never existed during the last 200,000 years, and most probably not even in the last 1–2 million years. That is a rapid and irreversible disappearance of sea-ice, leaving the Antarctic and Greenland ice-sheets surrounded by open ocean. It should be stated that this could not cause any immediate rise of sea-level since the floating ice is in equilibrium with the sea surface. It is impossible to forecast if this would cause the Greenland ice-sheet to melt, but if it did it would take several thousand years and cause a slow rise of sea-level by about 6 m. In contrast, the Antarctic ice-sheet would probably grow in size. Any drastic reduction of the Arctic sea-ice would cause the northern hemisphere climatic zones to move poleward.

Can it be predicted when the CO_2 concentration might double? Rotty and Weinberg (1977) comment that if the use of fossil fuels continues to grow at the present rate of 4 per cent per year, the total CO_2 injected into the atmosphere by man since 1860 will reach 300 ppm in 40 years' time. Thus the atmospheric concentration of CO_2 would be increased, allowing for ocean storage, by 300 ppm by AD 2030, and the total concentration would be equal to the 600 ppm assumed in Figure 6.20. According to Keeling (1976) a 2·3 per cent per year continued increase in the use of fossil fuel would give 10 times the present annual CO_2 emissions one hundred years hence. This, according to Keeling's model (which includes effect of storage in the ocean and land biota), would result in an atmospheric concentration of CO_2 approximately 8 times the present, which is 4 times the concentration assumed in Manabe's model. The temperature rise is proportional to the logarithm of the CO_2 concentration: and eightfold increase therefore causes an average temperature rise that is 3 times the rise caused by a twofold increase in CO_2. Flohn (1977) states that even with cautious assumptions of the global growth rate of energy use (less than 4 per cent annually), an increase of the atmospheric CO_2 content by a factor 2·5–3 should be expected during the next century. If the use of coal as an energy source rises again in the future, and economic growth remains uncontrolled, the possibility of an increase of atmospheric CO_2 by a factor of even 5 (or more) after about 100 years cannot be excluded. Much depends on the rate of increase of fossil fuel use, since a shift to coal as a replacement for oil and gas gives much more carbon dioxide per unit of energy.

Flohn (1977) has suggested the following scenario for global climatic evolution during the next century. The century would begin with a climatic state similar to that found in the period 1931–60, which in comparison with the previous 500 years was quite exceptional. The second act, around the middle of the century (or perhaps in 50–60 years' time), would show a return to the warm epoch between 800 and 1200 AD or to the post-glacial optimum, most probably with a marked retreat of the permanent Arctic drift-ice. Under such conditions, the third alternative (a warm age) mentioned at the start of the chapter has a high probability after the mid-twenty-first century, and forms the third act of the scenario.

Figure 4.20 has enabled Flohn to estimate the northward displacement of the subtropical anticyclones using reasonable estimates of the tropospheric temperature above an ice-free Arctic Ocean which might occur in the twenty-first century. This yields a northward displacement of the sub-tropical anticyclones during winter by 300–600 km, and thus allows estimates of the shifts of all other climatic belts. Flohn suggests that one consequence might be a reduction of the subtropical winter rain belt, with catastrophic consequences for the water supply of California and Utah, of the whole Mediterranean, and Near East up to Pakistan and Russian Central Asia.

The second alternative for future climate suggested at the start of this chapter was a transition to an ice age. The arguments for this have been discussed in some detail by Calder (1974). Calder takes the view that we have been for several thousand years on the descending branch of the Milankovitch beat frequencies discussed in Chapter 2. He also suggests from the climatic proxy records that ice-caps form suddenly. For this he coins the term 'snowblitz', a rapid extension of the snow cover that will not melt in summer. Chapter 5 described the rather rapid transitions between glacial and interglacial and vice versa during the Pleistocene. Such transitions seem to be switches between two near-equilibrium states occurring on a time-scale of some 100 years. Some of these switches are too short-lived to produce a fully developed glaciation of the northern continents, which necessitates 5,000–8,000 years, but instead they form abortive glaciations with a time-scale of about 200–2,000 years. Flohn (1977) suggests that the probability of the occurrence of such a glaciation during the next century is of the order of $0 \cdot 1$–$1 \cdot 0$ per cent. In contrast he estimates the probability of a transition to a warm period to exceed 10 per cent and possibly reaching 50 per cent. He considers that a climatic evolution towards a warm period is more likely than a return to the 'Little Ice Age weather' of between 1550 and 1850.

Rotty and Weinberg state that if one examines predicted disasters—ozone depletion, catastrophic reactor accident, perhaps even nuclear war—one must admit that the serious climatic shifts caused by CO_2 may be at least as likely as these, and more likely than some of them. Perhaps the most disturbing aspect of the CO_2 problem is that the ocean reservoirs restore the CO_2 in the atmosphere to equilibrium only after hundreds of years. Thus there is the serious possibility of putting CO_2 into the atmosphere in amounts which are later discovered to cause undesirable climatic changes. Atmospheric pollution must take its place in the energy debate alongside such problems as the hazards of nuclear power, since it cannot be assumed that coal burning is even a safe short-term alternative to nuclear energy.

Possible climatic changes can influence mankind in the comparative short term. Schneider (1977) has made the point that society is dangerously vulnerable to natural climatic variability at times of depleted food reserves. It was suggested in Chapter 6 that the world is now entering a period when global food reserves will be small. Schneider comments that it is not clear that highly productive US farmers will be able to increase their productivity in the future at anywhere near the same rate that productivity increased during the 1950's and 1960's without technological breakthroughs that cannot now be guaranteed. This is important since the US is the principal grain-exporting nation in the world, and since US agricultural exports provide not only an important economic benefit to the US, but an essential backstop against famine elsewhere. Most of the years between 1956 and 1972 were highly favourable climatically for agricultural production, particularly in North America. The years ahead may not be so favourable and may lead to large variations in world food production.

Landsberg (1976) comments that cause and effect relationships of climate are known in only a rudimentary way; hence predictions for either short or long intervals are afflicted with great uncertainties. The above forecasts of future climate have therefore to be approached with extreme caution. The only safe conclusion is that considerably more research is required into the nature and causes of climatic variations.

References

CALDER, N. 1974: *The Weather Machine*. London: BBC.

DAMON, P. E. and KUNEN, S. M. 1976: Global cooling? *Science 193*, 447–53.

DANSGAARD, W., JOHNSON, S. J., CLAUSEN, N. B. and LANGWAY, C. C. JR 1971: Climatic record revealed by the Camp Century ice core. In TURAKIAN, K. K., editor, *The late cenozoic glacial ages*. Hew Haven: Yale University Press.

EDDY, J. A. 1976: The Maunder minimum. *Science 192*, 1189–202.

FLOIIN, II. 1977: Climate and energy; a scenario to a 21st Century problem. *Climatic Change 1*, 5–20.

KEELING, C. D. 1976: Impact of industrial gases on climate. In *Energy and climate: outer limits to growth*. (National Academy of Science.)

KONDRATYEV, K. YA., and NIKOLSKY, G. A. 1970: Solar radiation and solar activity. *Quarterly Journal Royal Meteorological Society 96*, 509–22.

LAMB, H. H. 1970: Volcanic dust in the atmosphere; with a chronology and assessment of its meteorological significance. *Philosophical Transactions Royal Society Series A 226*, 425–533.

LANDSBERG, H. E. 1976: Whence global climate; hot or cold? an essay review. *Bulletin American Meteorological Society 57*, 441–3.

MANABE, S. and WETHERALD, R. T. 1975: The effects of doubling the CO_2 concentration on the climate of a general circulation model. *Journal Atmospheric Science 32*, 3–15.

MILANKOVITCH, M. 1930: Mathematische Klimalehre und astronomische theorie der Klimaschwankungen. In KÖPPEN, W. and GEIGER, R., editors, *Handbuch der Klimatologie 1*, Berlin: Teil A.

MITCHELL, J. M. JR 1970: A preliminary evaluation of atmospheric pollution as a cause of the global temperature fluctuation of the past century. In SINGER, S. F., editor, *Global effects of environmental pollution*, 97–112. New York: Springer.

ROTTY, R. M. and WEINBERG, A. M. 1977: How long is coal's future? *Climatic Change 1*, 45–57.

SCHNEIDER, S. H. 1977: Climate change and the world predicament. *Climatic Change 1*, 21–43.

SCHNEIDER, S. H. and MASS, C. 1975a: Volcanic dust, sunspots and temperature trends. *Science 190*, 741–6.

1975b: Volcanic dust, sunspots and long-term climate trends: theories in search of verification. In *Proceedings of the WMO/IAMAP symposium on long-term climatic fluctuations*, 365–9. Geneva: WMO.

Further reading

In addition to the references given at the end of the chapters the following books will be found to provide useful additional reading.

BRUCE, J. P. and CLARK, R. H. 1966: *Introduction to hydrometeorology*. Oxford: Pergamon Press.

FLOHN, H. 1969: *Climate and Weather*. London: Weidenfeld and Nicolson.

GEIGER, R. 1965: *The climate near the ground*. Cambridge, Mass.: Harvard University Press.

LAMB, H. H. 1972: *Climate: present, past and future* **1**, *Fundamentals and climate now*. London: Methuen.

1977: *Climate: present, past and future* **2**, *Climatic history and the future*. London: Methuen and New York: Barnes and Noble.

LOCKWOOD, J. G. 1974: *World climatology: an environmental approach*. London: Edward Arnold.

PITTOCK, A. B., FRAKES, L. A., JENSSEN, D., PETERSON, J. A. and ZILLMAN, J. W. 1978: *Climatic change and variability: a southern perspective*. London: Cambridge University Press.

RIEHL, A. 1965: *Introduction to the atmosphere*. New York: McGraw-Hill.

SELLERS, W. D. 1965: *Physical climatology*. Chicago: University of Chicago Press.

Glossary

Included in the following glossary are a few technical terms which may be less familiar to some readers. For further details, readers should consult one of the published meteorological glossaries, the best of which is *Meteorological Glossary* published by HMSO, London, 1972 and compiled by Dr D. H. McIntosh for the Meteorological Office.

ABLATION: The disappearance of snow and ice by melting and evaporation.

ABSORPTION: Removal of radiation from an incident solar or terrestrial beam, with conversion to another form of energy—electrical, chemical or heat.

ADIABATIC: An adiabatic process (thermodynamic) is one in which heat does not enter or leave the system.

ADVECTION: The process of transfer of an air-mass property by virtue of motion. The term is often used to signify horizontal transfer only.

AEROSOL: In meteorology, an aggregation of solid or liquid minute particles suspended in the atmosphere.

AIR MASS: A body of air in which horizontal gradients of temperature and humidity are relatively slight and which is separated from an adjacent body of air by a more or less sharply defined transition zone (front) in which these gradients are relatively large (see also OPTICAL AIR MASS).

ALBEDO: a measure of the reflecting power of a surface, being that fraction of the incident radiation (total or monochromatic) which is reflected by a surface.

ANGULAR MOMENTUM: The angular momentum per unit mass of a body rotating about a fixed axis is the product of the linear velocity of the body and the perpendicular distance of the body from the axis of rotation.

ANTICYCLONE: That atmospheric pressure distribution in which there is a high central pressure relative to the surroundings. It is characterized on a weather chart by a system of closed isobars, generally approximately circular or oval in form, enclosing the central high pressure.

ATTENUATION: The depletion of electromagnetic energy (e.g. solar radiation, radio waves, etc.) which is effected by the earth's atmosphere and its constituents.

BLOCKING: The term applied in middle-latitude synoptic meteorology to the situation in which there is interruption of the normal eastward movement of depressions, troughs, anticyclones and ridges for at least a few days. A blocking situation is dominated by an anticyclone whose circulation extends to the high troposphere. The zonal circulation to the west is transferred into a meridional circulation branching polewards and equatorwards.

BOUNDARY LAYER: That layer of a fluid adjacent to a physical boundary in which the fluid motion is much affected by the boundary and has a mean velocity less than the free-stream value.

BOWEN RATIO: The ratio of the amount of sensible heat to that of latent heat lost by a surface to the atmosphere by the processes of conduction and turbulence.

CALORIE (or gram-calorie): A unit of heat, being the heat required to raise the temperature of 1 g of water by 1 degree C. It was decided at the Ninth General Conference of Weights and Measures (1948) that the joule (J) should replace the calorie (cal) as the unit of heat:

$$1 \text{ cal} = 4\cdot1855 \text{ J}$$

CAPILLARY POTENTIAL: A concept used in soil moisture studies, being the force of attraction exerted by soils on contained water, or the equivalent force required to extract the water from the soil against the capillary forces (surface tension) acting in the soil pores. It is generally expressed in the pressure unit of atmospheres.

CLOSED SYSTEM: A closed system is one in which there is no exchange of matter between the system and its environment though there is, in general, exchange of energy. The atmosphere as a whole may, to a high degree of approximation, be considered a closed system.

CONDUCTION (of heat): The process of heat transfer through matter by molecular impact from regions of high temperature to regions of low temperature without transfer of the matter itself. It is the process by which heat passes through solids; its effects in fluids are usually negligible in comparison with those of convection.

CONVECTION: A mode of heat transfer within a fluid, involving the movement of substantial volumes of the substances concerned. The convection process frequently operates in the atmosphere and is of fundamental importance in effecting the vertical exchange of heat and other air-mass properties.

CORIOLIS ACCELERATION: An apparent acceleration which the air possesses by virtue of the earth's rotation, with respect to axes fixed in the earth.

CORIOLIS PARAMETER: A quantity, denoted f, defined by the equation $f = 2\Omega \sin \varphi$, where Ω is the magnitude of the earth's angular velocity and φ the latitude.

DENDROCHRONOLOGY: The interpretation of the varying width of the annual growth rings of certain trees in terms of the corresponding year-to-year climatic fluctuations.

DIFFUSE RADIATION: Radiation which is received simultaneously from very many directions.

DIRECT CIRCULATION: A circulation in which potential energy represented by the juxtaposition of relatively dense and light air masses is converted into kinetic energy as the lighter air rises and the denser air sinks.

EVAPOTRANSPIRATION: The combined processes of evaporation from the earth's surface and transpiration from vegetation. 'Potential evapotranspiration' is the addition of water vapour to the atmosphere which would take place by these processes from a surface covered by green vegetation if there were no lack of available water.

FIELD CAPACITY: The mass of water (per cent of dry soil) retained by a previously saturated soil when free drainage has ceased is known as the soil's field capacity or water-holding capacity.

GENERAL CIRCULATION: The term 'general circulation' has different meanings in different contexts and there is no unique definition. In its widest sense it is used to imply all aspects of the three-dimentional global flow and energetics of the whole atmosphere.

GEOSTROPHIC WIND: That horizontal equilibrium wind (V_G), blowing parallel to the isobars, which represents an exact balance between the horizontal pressure gradient force and the horizontal component of the Coriolis force ($f V_G$). Low pressure is to the left of the wind vector in the northern hemisphere, to the right in the southern.

GLOBAL RADIATION: The sum of direct and diffuse radiation received by unit horizontal surface.

GREENHOUSE EFFECT: The effect whereby the earth's surface is maintained at a much higher temperature than the temperature appropriate to balance conditions with the solar radiation reaching the earth's surface. The atmospheric gases are almost transparent to incoming solar radiation, but water vapour and carbon dioxide in the atmosphere strongly absorb infrared radiation emitted from the earth's surface and re-emit the radiation, in part downwards.

HADLEY CELL: A simple thermal circulation first suggested by George Hadley in the 18th century as part explanation of the trade winds and still to be approximated to in the troposphere between latitudes 0° and 30°. If the effects of the earth's rotation are neglected, the circulation comprises upward motion over the equator, downward motion in the subtropics, poleward motion at high levels, and equatorward motion at low levels.

ICE-SHEET: A large area of land-ice with a dome-shaped, almost smooth surface. The largest ice-sheets now existing are those in Antarctica and Greenland.

INDEX CYCLE: A term often applied to alternating periods of predominantly zonal and meridional flow.

JET STREAM: A fast narrow current of air, generally near the tropopause, characterized by strong vertical and lateral wind shears. A jet stream is usually some thousands of kilometres in length, hundreds of kilometres in width and some kilometres in depth.

LATENT HEAT: The quantity of heat absorbed or emitted, without change of temperature, during a change of state of unit mass of a material. The latent heat of fusion (ice to water) at 0°C is 79·7 cal g^{-1}. The latent heat of vaporization (water to water vapour) at 0° C is 597 cal g^{-1}.

MERIDIONAL CIRCULATION: Generally, a closed circulation in a vertical plane oriented along a geographic meridian, i.e. Hadley cell.

MERIDIONAL FLOW: Airflow in the direction of the geographic meridian, south–north or north–south flow.

MIE SCATTERING: Scattering of electromagnetic radiation by spherical particles, much of the radiation being scattered in a forward direction.

NET RADIATION: The sum of the upward and downward components of the solar and the terrestrial radiation through a horizontal surface. Radiation fluxes directed downwards are considered to be positive while those directed upwards are negative.

OPEN SYSTEM: An open system is one in which there is an exchange of energy and matter between the system and its environment.

OPTICAL AIR MASS: The length of the path of the sun's rays through the earth's atmosphere, measured in terms of the path length when the sun is in the zenith (overhead position).

PERMAFROST: Soil which remains permanently frozen, summer heating being insufficient to raise above freezing point the temperature of the lower part of a frozen layer formed during winter.

POTENTIAL ENERGY: The energy possessed by a body by virtue of its position. It is measured by the amount of work required to bring the body from a standard position, where its potential energy is zero, to its present position. A common example is that of 'gravitational potential energy', mean sea-level being then the normal selected standard level.

RAYLEIGH SCATTERING: Scattering of electromagnetic radiation affected by spherical particles of radius less than about one-tenth the wavelength of the incident radiation. The Rayleigh scattering of incident solar radiation by air molecules explains the blue colour of the sky.

SATURATION DEFICIT: The difference between the actual vapour pressure of a moist air sample at a given temperature and the saturation vapour pressure corresponding to that temperature.

SENSIBLE HEAT: The transfer of sensible heat to a body causes an increase in the temperature of the body. See also latent heat.

SOIL MOISTURE DEFICIT: The amount of rainfall required to restore soil to its field capacity.

SPECIFIC HEAT: The specific heat of a substance is the heat required to raise the temperature of unit mass of it by one degree.

SPECIFIC HUMIDITY: The ratio of the mass of water vapour to the mass of moist air in which it is contained.

TEMPERATURE: The condition which determines the flow of heat from one substance to another.

TEMPERATURE SCALES: The scales in common climatological use are the Celsius (or centigrade) scale, and the Kelvin (or absolute) scale. On the Celsius scale the freezing and boiling points of water at standard pressure are respectively 0 °C and 100 °C. Kelvin is the appropriate scale to use in basic physical equations which involve temperature. In meteorological practice $K = 273 + °C$ with sufficient accuracy.

THICKNESS: The height difference at a given place between specified pressure levels, usually 1,000 and 500 mbar.

TURBIDITY: That property of a cloudless atmosphere which produces attenuation of solar radiation. Measurements of atmospheric turbidity are generally concerned with the attenuation which is additional to that associated with molecular scattering, the particles responsible being mainly dust and smoke.

WILTING POINT: The point at which the soil contains so little water that it is unable to supply it at a rate sufficient to prevent permanent wilting of plants.

ZONAL FLOW: West–east airflow. East–west airflow is generally reckoned as negative zonal flow.

Subject Index

Author Index